STUDENT WORKBOOK

for

INTERPERSONAL PROCESS IN PSYCHOTHERAPY

A RELATIONAL APPROACH

Fourth Edition

Edward Teyber

with

Faith H. McClure

Brooks/Cole
Thomson Learning™

Australia • Canada • Mexico • Singapore • Spain • United Kingdom • United States

ISBN 0-534-36566-3

For more information, contact
Wadsworth/Thomson Learning
10 Davis Drive
Belmont, CA 94002-3098
USA
http://www.wadsworth.com

International Headquarters
Thomson Learning
International Division
290 Harbor Drive, 2nd Floor
Stamford, CT 06902-7477
USA

UK/Europe/Middle East/South Africa
Thomson Learning
Berkshire House
168-173 High Holborn
London WC1V 7AA
United Kingdom

Asia
Thomson Learning
60 Albert Complex, #15-01
Singapore 189969

Canada
Nelson Thomson Learning
1120 Birchmount Road
Toronto, Ontario M1K 5G4
Canada

TABLE OF CONTENTS

i

CHAPTER FIVE: RESPONDING TO CONFLICTED EMOTIONS

CHAPTER SIX: FAMILIAL AND DEVELOPMENTAL FACTORS

CHAPTER SEVEN: INFLEXIBLE INTERPERSONAL COPING STYLES

CHAPTER EIGHT: CURRENT INTERPERSONAL FACTORS

CHAPTER NINE: AN INTERPERSONAL SOLUTION

CHAPTER TEN: WORKING THROUGH AND TERMINATION

INTRODUCTION TO THE STUDENT WORKBOOK

<u>Interpersonal Process in Psychotherapy: A Relational</u>
<u>Approach</u> (4/E) is composed of ten chapters. This Student
Workbook is divided into four parts for each chapter of the
text. The workbook provides students with a detailed Study
Guide, Self-Reflection Exercises, Supportive Readings and Case
Studies, and space for Journal Entry. For use in conjunction
with the text, the workbook is designed to help students: (1)
study for classroom examinations; (2) integrate the personally
evocative and affectively laden material presented in the text;
(3) examine important supplements that further develop key
chapter concepts; and (4) explore their personal reactions to
clinical training and working with the *Interpersonal Process*.

Part I, the Study Guide, consists of a list of Key Terms,
thought provoking Essay questions to help students grasp the
central principles, and sample Multiple Choice and True/False
questions to help students focus on essential constructs. This
section will prepare students for course examinations.

Part II, Self-Reflections, consists of personally evocative
questions that help students relate the conceptual material in
the text to their own personal experience and to the lives of
their clients. The questions are also designed to help students
cope more effectively with the challenges of intensive clinical
training. As the need for further support or understanding

arises, trainees can discuss issues related to these exercises with a trusted colleague, supervisor, or therapist.

Part III, Supportive Readings and Case Studies, provides important new material that further develops central concepts discussed in each chapter of the text. Most of the case studies are structured using the Case Conceptualization/Treatment Planning format presented in Appendix B. The Case Studies are based on characters from well-known films and plays (i.e., "Death of a Salesman" and "Ordinary People"), as well as from the case files of skilled therapists. These Case Studies provide a model to help students see how to apply the Case Conceptualization/Treatment Planning format to actual cases. Students will be able to use these materials to better integrate and apply the concepts presented in the text.

Part IV, Journal Entry, allows students to explore their personal reactions to the material presented both in the text and the workbook. The space can also be used to note questions or issues that arise and need to be addressed during practicum or class time.

There is more information to work with in each of these chapters than students can address. It may be best if you can approach these exercises as options to explore rather than assignments to complete. One useful strategy is to peruse each chapter cursorily and then return selectively to a few issues or

questions that were most salient for you. Finally, interested
readers are invited to make editorial comments or other
suggestions for future editions of this workbook. Please e-mail
your comments to: eteyber@csusb.edu.

CHAPTER ONE: INTRODUCTION AND OVERVIEW

PART I: STUDY GUIDE

Part I will help students prepare for course examinations. It consists of three sections: a list of key terms to define, essay questions for central concepts, and sample multiple choice and true/false questions.

Section A: Key Terms

Provide a one or two sentence answer for the following key terms.

RELATIONAL VS DRIVE MODELS_____

_____

_____

ROLE OF ANXIETY ACCORDING TO INTERPERSONAL THEORY_____

_____

_____

SELF-SYSTEM_____

_____

_____

TRANSACTIONAL PATTERNS _____

_____

_____

INTERPERSONAL COPING STYLES/STRATEGIES _____

_____

_____

OBJECT/CONSTANCY _____

_____

_____

SCHEMAS, TEMPLATES, OR INTERNAL WORKING MODELS _____

_____

_____

SPLITTING DEFENSES _____

_____

_____

SCAPEGOAT _____

_____

_____

SEPARATE-RELATEDNESS CONTINUUM _____

_____

_____

STRAIN TRAUMA VS SHOCK TRAUMA _____

_____

_____

PROCESS VS CONTENT _____

_____

_____

RECAPITULATION _____

_____

_____

CORRECTIVE EMOTIONAL EXPERIENCE _____

_____

_____

INTERPERSONAL SAFETY _____

_____

_____

TRANSFERENCE TESTS _____

_____

_____

CLIENT RESPONSE SPECIFICITY _____

_____

_____

SUBJECTIVE WORLDVIEW _____

_____

_____

FAMILIAL ROLES _____

_____

_____

Section B: Key Concepts

Consider each of the following key concepts. Be prepared to write a two to three paragraph essay for each of the following questions.

1. The "corrective emotional experience" is a cardinal concept in the interpersonal process approach. Discuss how client conflicts may be reenacted versus resolved in the therapeutic relationship.

_____

_____

_____

_____

_____

_____

_____

_____

2.   Discuss the importance of the therapist's personal flexibility in responding to client diversity. Discuss how diversity and "client response specificity" are related. Illustrate with an example.

_____

_____

_____

_____

_____

_____

3.   The interpersonal process approach is based on three
      theoretical systems: (1) interpersonal, (2) object
      relations, and (3) family systems. Discuss the <u>most</u>
      relevant contributions from each of these systems to this
      relational approach.

_____

_____

_____

_____

_____

_____

_____

_____

_____

Section C: Sample Multiple Choice and True/False Questions

    Provide answers to the following sample multiple choice and
true/false questions.

1.   According to Sullivan's theory, _____ is a collection
      of interpersonal strategies employed to avoid or minimize
      anxiety, ward off disapproval, and maintain self-esteem.
      a.   Mental Illness
      b.   Corrective Emotional Experience
      c.   Personality
      d.   Behavioral Adaptation

2.   Which theory defines the primary motivation in individuals
      to establish and maintain secure emotional ties to
      parental caregivers?
      a.   Freudian
      b.   Family Systems
      c.   Cognitive Behavioral
      d.   Object Relations

3.   Patterns of interaction and communication that become
      repetitive and rule-bound are developed for the purpose of:
      a.   Homeostasis.
      b.   Splitting.
      c.   Diffusion.
      d.   Rigidity.

4.   Therapists that utilize the interpersonal process approach
      must shift entirely away from the specific content of what
      the client is discussing and focus solely on the relational
      process.
      a.   true
      b.   false

5.   The cornerstone of change according to the interpersonal
      process approach is to provide clients with an
      "experience" rather than an "explanation."
      a.   true
      b.   false

      Answers: 1.c, 2.d, 3.a, 4.b, 5.a

PART II: SELF-REFLECTIONS

The material presented in the text is often personally evocative and challenging. The questions below are designed to help students process and integrate their reactions to this information.

1.   In Chapter One we discussed how the therapist needs to: (1) be able to establish and maintain the therapeutic alliance, (2) conceptualize clients' dynamics and provide a focus for treatment, and (3) enter clients' subjective worldview and respond to their feelings. At this point in your clinical training, how do you evaluate your ability in each of these three arenas? We will return to this question in Chapter Ten and see how you assess yourself on each of these dimensions later in your development.

_____

_____

_____

_____

_____

_____

_____

_____

_____

_____

_____

_____

_____

2.   At this stage in your clinical training, what is the
     primary source of anxiety or insecurity that may be evoked
     in you by seeing clients?

_____

_____

_____

_____

_____

_____

_____

_____

_____

_____

_____

_____

_____

_____

_____

_____

3.   Continuing from question number 2, can you
identify the types of responses from supervisors,
clients, or others that could exacerbate these
concerns, as well as how others could respond to help
alleviate them?

_____

_____

_____

_____

_____

_____

_____

_____

_____

_____

_____

_____

_____

_____

_____

_____

_____

4.   What is, or what do you anticipate to be, the primary
      satisfaction or personal reward in being a counselor?

_____

_____

_____

_____

_____

_____

_____

_____

_____

_____

_____

_____

_____

_____

_____

_____

_____

PART III: SUPPORTIVE READINGS AND CASE STUDIES

To respond effectively to a client's emotions and inner life, the counselor must identify unifying patterns in personality, as well as take into account family and cultural features that contribute to the client's problems. The following case example illustrates the concept of "client response specificity" by giving the client a corrective emotional experience within the context of the client's multicultural background.

Section A: The Therapeutic Process: Case Study of Teresa

Two of the most important concepts in Chapter One are distinguishing the process level from the content level, and effecting change by providing a corrective or reparative experience in the interaction or real-life relationship between the counselor and client. The case vignette below helps to clarify both of these cardinal concepts which we will be examining throughout this text.

Suppose a counseling trainee and her 17-year-old client, Teresa, are talking for the first time about Teresa's sexual contact with her stepfather. The content of what they are talking about is sexual molestation. Depending on the process they enact, however, the effectiveness of this discussion will

vary greatly. On the one hand, suppose the counselor is
initiating this discussion and pressing Teresa for disclosure
about what occurred. The therapist is a graduate student who is
genuinely concerned about Teresa's safety. Her "need to know"
is intensified by her concerns about her legal responsibilities
as a mandated reporter, and her concern that her supervisor will
want her to know more details or "facts" about Teresa's
molestation.

In response to the counselor's continuing press for
disclosure, Teresa complies with the therapist's authority and
reluctantly speaks. Although useful information may be gained
under these circumstances, the opportunity for meaningful
therapeutic gain is lost because aspects of Teresa's core
conflict are being reenacted with the counselor in the way they
are interacting. That is, Teresa is again being pressured to
obey an adult, comply with authority, and do something she
doesn't want to do. Of course, being pressured to talk about
something she doesn't want to disclose in no way retraumatizes
her as the original abuse did. However, their process of
demand/comply is awry, and it is likely to re-evoke in Teresa
the same core affective concerns the abuse initially engendered.
That is, her helplessness will lead to depression and her
compliance to feelings of shame. Because the interpersonal
process with the therapist is thematically re-evoking the

original conflict, her disclosure will hinder her progress in therapy and slow the process of re-empowerment.

The situation may be further complicated if Teresa belongs to an ethnic group where family loyalty is highly prized and if she is a member of a religious community where obedience to authority and hierarchical relationships are emphasized. In this situation, Teresa is being asked to violate rules sanctioned by her family's cultural group (i.e., religious, ethnic). Other family members and friends may not approve of the stepfather's behavior. Even so, they may not provide Teresa with the support she needs because she took this information to someone outside the family (the counselor), or because she "violated" their religious proscriptions that emphasize obedience and forgiveness. Thus, although disclosure violates rules and loyalties for most victims, this may take on additional significance and add much more distress for Teresa if she is a Hispanic adolescent who belongs to a conservative religious community.

What should the counselor in this example do instead? Wait nondirectively for Teresa to volunteer this information -- while she may continue to suffer ongoing abuse at home? Of course not. But by attending to the process dimension, the counselor may be able to begin providing Teresa with a reparative therapeutic experience while gathering the same information.

That is, instead of pressing for disclosure, what if the counselor *honored* Teresa's "resistance" or cultural prescriptions. For example, instead of pushing for more disclosure, the counselor could "metacommunicate" and make a "process" comment (Cashdan, 1988; Kiesler, 1996; Strup & Binder, 1984) by inquiring supportively about Teresa's reluctance to speak:

Counselor: "You do not want to talk with me about this right now. Something doesn't feel safe. I'm wondering what might happen at home if you talk with me about this?"

Or: "I'm wondering what kinds of things might go wrong between us if you talked to me about this?"

Or: "Can we talk instead about what you are concerned your mother or stepfather might think or do?"

Or: "Would you get more support or feel safer if you discussed this with your minister or another family member?"

Only if the counselor can help Teresa identify and resolve her potential concerns about disclosing can she find it relieving. After this reparative relational experience (i.e., Teresa's expectations of having to comply with authorities is not reenacted, and new respect for her limits and compassion for her experience is provided), Teresa will feel *safer* to engage with the counselor in discussing what happened and begin to make progress in treatment. Given her previous experience and expectations, there may certainly be "reality" to the concerns she presents. However, the following concerns typically need to be explored and clarified *before* Teresa can find the safety she needs to act on her own behalf, rather than continuing to comply as she has been doing in other problematic relationships. For example, in response to the questions suggested above, Teresa might reply that her mother won't believe her; that her stepfather will be sent away; she will be told it is all her fault; she will be chastised for not keeping it in the family and trying to resolve it there; that the counselor may take her parents' side; the counselor may want to remove her from her home immediately; the counselor may see her as "defective"; or, the counselor may feel uncomfortable with or inadequate to respond to this sensitive material and will not want to explore it fully.

Although this process difference may seem subtle, its effect is powerful and will have a far-reaching impact on the course and outcome of treatment. In the second scenario, Teresa is already beginning the empowerment process just by the way in which she shares her trauma. *For the first time in her life, she is able to say "no" or set limits with an authority figure and still remain emotionally connected and supported.* This reparative interpersonal process with the counselor is a prerequisite for the client to be able to begin acting in

similarly empowered, authentic, and self-affirming ways in some
other relationships. The counselor can then help Teresa to
systematically begin discerning other people in her life with
whom it is safe to be similarly assertive, and those who will
punish this or also demand that she comply.

        What were your thoughts and reactions to this case study?
If Teresa were your client, what would you do similarly and
differently?

_____

_____

_____

_____

_____

_____

_____

_____

_____

_____

_____

_____

PART IV: JOURNAL ENTRY

Use the following section to integrate the material presented, to address unresolved questions or issues, and to further explore your personal reactions.

_____

_____

_____

_____

_____

_____

_____

_____

_____

_____

_____

_____

CHAPTER TWO: ESTABLISHING A COLLABORATIVE RELATIONSHIP

PART I: STUDY GUIDE

Part I will help students prepare for course examinations. It consists of three sections: a list of key terms to define; essay questions for central concepts emphasized in the text; and sample multiple choice and true/false questions.

Section A: Key Terms

Provide a one to two sentence answer for the following key terms.

INVALIDATION/MYSTIFICATION _____

_____

_____

EFFICACY _____

_____

_____

WORKING ALLIANCE/COLLABORATIVE RELATIONSHIP _____

_____

_____

HOLDING ENVIRONMENT _____

_____

_____

PRESENCE _____

_____

_____

CORE CONDITIONS _____

_____

_____

DIRECTIVE-NONDIRECTIVE BALANCE _____

_____

_____

CLIENT COMPLIANCE_____

_____

_____

TRIANGULATION _____

_____

_____

PROCESS COMMENTS _____

_____

_____

COGNITIVE FLEXIBILITY AND PERSPECTIVE TAKING _____

_____

_____

DEMONSTRATING UNDERSTANDING _____

_____

_____

ATTUNED RESPONSIVENESS _____

_____

_____

EMPATHY _____

_____

_____

"GOOD ENOUGH" MOTHERING _____

_____

_____

CORE MEANING/EMOTIONAL MESSAGE _____

_____

_____

REENACTING DEVELOPMENTAL CONFLICTS _____

_____

_____

TOLERATING AMBIGUITY _____

_____

_____

INTEGRATING FOCUS _____

_____

_____

REPETITIVE RELATIONAL PATTERNS _____

_____

_____

PATHOGENIC BELIEFS _____

_____

_____

COGNITIVE TRIAD _____

_____

_____

HOT COGNITIONS _____

_____

_____

PATHOGENIC BELIEFS AND FAULTY EXPECTATIONS _____

_____

_____

RECURRENT AFFECTS _____

_____

_____

IMMEDIACY _____

_____

_____

THERAPEUTIC IMPACT DISCLOSURE _____

_____

_____

METACOMMUNICATIVE FEEDBACK _____

_____

_____

SELF-INVOLVING STATEMENTS _____

_____

_____

COUNTERTRANSFERENCE _____

_____

_____

PERFORMANCE ANXIETIES _____

_____

_____

THIRD EAR _____

_____

_____

Section B: Key Concepts

Consider each of the following key concepts. Be prepared to write a two to three paragraph essay for each of the following questions.

1.   You have been assigned a new client. John is a depressed 27 year-old male who is experiencing symptoms of anxiety and depression. He lives alone, works 24 hours a week as a waiter, and is a full-time student at the local university, where he is majoring in chemistry. John is having trouble concentrating and motivating himself to study. Throughout this session, John is having difficulty in initiating. Prior to seeing John, you received an intake report that mentioned that John's brother recently died in a gangland shooting. John states that his parents considered his brother to be perfect and that no matter how he tried John was always getting in trouble at home. Formulate two or three working hypotheses you would generate to prepare yourself to better establish a collaborative relationship with John.

_____

_____

_____

_____

_____

_____

_____

_____

_____

_____

_____

2.   What is meant by "finding an integrating focus" for
      treatment? How does the therapist begin to identify a
      treatment focus, and how might the therapist provide this
      focus in a collaborative manner?

      _____

      _____

      _____

      _____

      _____

      _____

      _____

3.   Describe the importance of balancing "directive" and "non-
      directive" approaches in establishing a working alliance
      with clients. How does this relate to "client response
      specificity" and the "separateness-relatedness dialectic?

      _____

      _____

      _____

      _____

      _____

      _____

      _____

      _____

4. Discuss reasons why therapists tend to avoid the unspoken
    emotional messages in clients' narratives rather than
    utilizing their "third ear." Elucidate how this avoidance
    may tend to reenact rather than resolve clients' core
    conflicts.

    _____

    _____

    _____

    _____

    _____

    _____

    _____

    _____

5. Discuss the importance of understanding clients' subjective
    worldview. Include possible sources of misunderstanding
    and disconfirmation. What can therapists do to ensure they
    accurately enter their clients' worldview? How can
    therapists demonstrate their understanding of their
    clients?

    _____

    _____

    _____

    _____

    _____

    _____

    _____

    _____

6.   Explain the difference between self-involving statements and self-disclosing statements. How might they facilitate and/or impede the therapeutic relationship?

_____

_____

_____

_____

_____

_____

_____

_____

_____

Section C: Sample Multiple Choice and True/False Questions

Answer the following sample multiple choice and true/false questions.

1.   Pervasive feelings of disempowerment or inefficacy are a consequence of (choose the <u>best</u> response):
     a.   childhood trauma.
     b.   systematic invalidation.
     c.   mental illness.
     d.   "good enough" mothering.

2.   Identifying recurrent themes will help beginning therapists:
     a.   obtain an interactive focus about the content of the client's problem.
     b.   obtain important information about the client's ability to tell the truth.
     c.   obtain an integrating treatment focus.
     d.   feel more at ease with their clients.

3.   During the initial session, the therapist must make an overt bid to establish a collaborative relationship.
     a.   true
     b.   false

Answers: 1.b, 2.c, 3.a

PART II: SELF-REFLECTIONS

The material presented in the text is often personally evocative and challenging. The questions below are designed to help students process and integrate their reactions to this information.

1.   Recall a significant crisis in your life that occurred within the last three years. Consider effective and ineffective responses that you received from significant others. What did others say and do that was helpful to you, and what did others say and do that was not? How did each type of response make you feel?

_____

_____

_____

_____

_____

_____

_____

_____

_____

_____

_____

_____

_____

_____

2.   How do you _typically_ or initially respond to the crisis
     events and emotional needs of significant others in your
     life? When others are hurting or in pain, what do you
     usually feel, and what do you tend to say and do? In what
     ways has your _new role_ as a graduate student therapist had
     an effect on your responses to family members or friends?
     Have you noticed a change in your interpersonal
     relationships?

_____

_____

_____

_____

_____

_____

_____

_____

_____

_____

_____

_____

_____

_____

_____

_____

PART III: SUPPORTIVE READINGS AND CASE STUDIES

It is often challenging for a student therapist to find an integrating focus for the wide diversity of information that clients present. Drawing on attachment theory and the related concepts of templates, internal working models, or cognitive schemas, the therapist can identify repetitive relational patterns that are central to clients' problems and distress. The following excerpt provides useful definitions and clarifying illustrations of these concepts. (Reprinted from Colin, V. (1996). *Human attachment* (pp. 19-22). New York: McGraw Hill.

Section A: Understanding Relational Patterns (V. Colin, 1996)

REPRESENTATIONAL MODELS

It is possible to assess and describe a baby's attachments by observing, describing, and making inferences from the baby's behavior. After infancy, however, you cannot describe an attachment adequately without describing the individual's representational models of relationships. A representational model, or "working model," of an attachment relationship is a mental representation of the self, the other, and their relations. Representational models include feelings, beliefs, expectations, behavioral strategies, and rules for directing attention, interpreting information, and organizing memory.

For example, consider Doug's mental model of himself. Doug is 24 years old. He feels secure. He believes that he is lovable and competent. He expects to succeed in his endeavors. He ordinarily employs a strategy of communicating openly and directly with others when he needs practical assistance or emotional support. However, he recognizes that some of the people he knows cannot be trusted to respond helpfully, so he makes intelligent choices about where to seek assistance. He notices when things do not go as expected and enjoys analyzing feedback from others. His mental model includes a rule that encourages making comparisons of and contrasts between bits of data for the purpose of integrating them into an accurate, coherent understanding of reality. He can remember incidents and qualities of relationships from his childhood easily and accurately.

For the last 3 years, Doug has been living with John, who has become one of the major attachment figures in Doug's life. In times of stress, such as when he lost his job, Doug seeks support and care from John more often than he seeks it from his parents. Doug trusts John to treat him well. His representational model of their relationship includes the expectation that John will be available and responsive when called upon.

Now consider Bill's representational models. Bill is 24 years old and has been married for 3 years. He seldom acknowledges feeling anxious, but the people who know him well say he seems insecure and hostile. He thinks it is important to be able to handle himself "like a man" in any situation that comes up. He believes that he must rely on himself to accomplish his goals and meet life's challenges. His preferred strategy is to figure things out on his own. He does sometimes seek information from others, but he avoids emotional interactions, even with his wife. He hates needing support or care from anyone and does not expect support or care to be available if he ever does need it. He does not like to think about his own nature, about others' personalities, or about relationships. He finds such matters confusing and distasteful. Bill thinks his parents were wonderful but does not remember many specific experiences from his childhood. He does not think what happened when he was a child is very important anyway.

Bill is annoyed by his wife's wanting to talk about her feelings and his. He finds her unpleasantly intrusive and clingy. However, he is vigilant for signs of disloyalty or independence. He wants her to be home always when he gets there and not to get sexually or emotionally involved with anyone else. He is not sure he can trust her. Bill's representational model of what attachment relationships are generally like is anxious.

If your models do represent reality with a fair degree of accuracy, then being able to use memory, information, and cognitive skills to choose a course of action saves time and effort. However, representational models do not always match reality well. Instead, the mental models an individual has developed can constrict or distort information processing and decision making. Consider Bill's models, described above. It is quite likely that many of the people with whom Bill interacts would be willing to work with him cooperatively an even to provide emotional support if he would only permit it. His ingrained distrust of others and his compulsive self-reliance prevent him from discovering those aspects of reality. His avoidance of emotional intimacy prevents him from developing a secure, trusting relationship even with his wife. In Bill's case, the distortions that result from his representational models are systematic. In other cases, distortions may be chaotic.

HOW REPRESENTATIONAL MODELS PERPETUATE EARLY PATTERNS

Once formed, representational models tend to maintain their coherence and pattern of organization. The individual acquires some fairly stable personality tendencies. New social partners are selected on the basis of, and/or assimilated to, old models

of people and relationships. As defined by Piaget (1970),
assimilation is the process of incorporating a new object into
an existing mental representation.

Just when and how internal models form, consolidate, and
become increasingly self-perpetuating are matters of some
uncertainty. Recently, attachment theorists have been delving
into new theoretical models developed by cognitive psychologists
and studying the insights thus obtained.

According to attachment theory, the individual's
representational models set the stage for interactions with new
social partners and have long-lasting consequences for
personality development and for the nature of close
relationships. The pattern of attachment behavior and the
associated representational models of attachment relationships
are adaptive for the environment in which the young child
develops them. Each child does the best she can to get the
protection and care every child needs, but she does it with
limited physical and cognitive abilities, and she does it in the
context of the qualities of care her family provides or
arranges. From her actual experiences, she develops
representational models of herself and of attachment
relationships. Formed unconsciously and early in childhood,
these models may constrain flexibility later and preclude
optimal adaptations to later circumstances and possibilities.

For example, a child whose mother resents the burdens and constraints of caring for him and frequently rebuffs his approaches is likely to develop a representational model of attachment that includes the knowledge that being open and direct about his needs will not work. He may also infer that he is unlovable, unworthy, and incompetent. He is likely to remain emotionally needy and angry.

Consider this example. When a little boy showed his mother his drawing, she was too busy to look. When, 5 hours after breakfast, he pestered her for lunch, she ignored his pleading. When his little sister knocked down his block tower and threw up on his teddy bear, the mother offered him no comfort and made him help clean up the mess. When he cried for attention, she scolded him and sent him to his room alone. Only when he spoke softly to her, brought her a cookie, and caressed her cheek, did she finally let him sit for a moment in her lap and lean against her. These sorts of experiences were repeated day after day, month after month. (Almost all mothers have a bad day once in a while, and it doesn't ruin a child's life.)

For the boy in this example, the strategy that worked most often was appearing to meet the mother's needs as a way of getting her to meet his. The representational model derived from his childhood experiences will probably lead him to be distrustful, manipulative, and resentful in new attachment relationships. If he bases his perceptions and behavior in new relationships on the model developed in his first attachment relationship, he may never discover that some people are quite willing to give nurturance and protection when he simply asks for it.

MULTIPLE MODELS

One more proposition about representational models must be introduced here. Bowlby (1980) asserted that it is not uncommon for a person to hold two conflicting internal models of an important relationship. This can occur when what a child is told contradicts her actual experience or when the experience itself is simply too painful to bear remembering. One of the child's representational models develops, Bowlby suggested, largely from direct experience, encoded and stored in episodic memory. A second, contradictory model may develop largely from cognitive input if what the parents tell the child conflicts with her actual experience.

For example, a child may discover repeatedly that when he is frightened or hurt and seeks contact with one of his parents the parent rebuffs his appeal and scolds him for crying. He is likely to learn to deny his distress, his need for comfort and protection, and his parent's rejection of him. "Denying" a feeling, an impulse, or a piece of information means blocking it from awareness, keeping it unconscious, fiercely pretending it does not exist and is not important. The suppression of such feelings and behavioral tendencies from conscious awareness may be fortified by the parent's statements that he loves the child, is proud of him, and gives him good care. Powerless to change the parent's behavior, the child can adapt to his circumstances best by suppressing awareness of some of the emotional realities of his situation. One mental model of the parent will include the awareness that the parent does not nurture or protect the child when he needs it. The emotions associated with that model will include fear and anger. That whole model and the associated feelings are likely to be repressed. A second representational model, the one that is likely to be conscious, will say that the child has a fine parent. Evidence of the repressed model is likely to appear in the child's behavior. When frightened by an unexpected separation from his parent, the child may seek to hide his anxiety and his anger even from himself, but he may hit or kick another child without apparent provocation or redirect his aggressive impulses to objects in the environment and heave or break something.

A child may also develop multiple, contradictory models of an attachment figure and of the associated attachment relationship if what happens to her is too painful to keep in awareness. This occurs often in cases of sexual abuse. For example, if the man to whom a child became attached in early childhood and on whom she still depends for some measure of protection, affection, and care uses her for his own sexual gratification, the child is likely to "split" her knowledge about her father into two isolated mental models. She will have one model of a more-or-less caring, protective father and another model of a sexual, demanding father who brings her terror and pain. Clinical evidence makes a persuasive case that the second model and associated memories may be entirely repressed. The girl may have no conscious knowledge of them for years. Some event in later life may trigger their reemergence as conscious memory. Or, in other cases, access to memories of the traumatic experiences may remain blocked indefinitely. A 50-year-old incest survivor with a kind husband may have no idea why she often experiences a panic attack (rapid breathing, rapid heartbeat, a feeling of terror) when her husband initiates sexual foreplay. A male survivor of sexual abuse may have no idea why he feels himself floating away from his body and experiences no physical sensations when a woman he finds attractive behaves seductively.

The blocking of memories of painful past experiences may prompt a person, years after the traumatic events have stopped, also to block awareness and processing of cues related to current, similar experiences. Sadly, this blocking may leave the victim unable to recognize and so to avoid victimization in a new relationship.

When an individual has conflicting models of a single relationship, he or she is likely to have limited access or even no conscious access to one or more of her or his representational models. As the examples above illustrate, the unconscious model(s) may nevertheless profoundly affect his or her personality and behavior.

1.   Describe one of your own internal working models or relational templates that created conflict in your adult relationships that you were able to identify and change.

_____

_____

_____

PART IV: JOURNAL ENTRY

Use the following section to integrate the material presented, to address unresolved questions or issues, and to further explore your personal reactions.

_____

_____

_____

_____

_____

_____

_____

_____

_____

_____

_____

_____

CHAPTER THREE: HONORING THE CLIENT'S RESISTANCE

PART I: STUDY GUIDE

Part I will help students prepare for course examinations. It consists of three sections: a list of key terms to define; essay questions for central concepts; and sample multiple choice and true/false questions.

Section A: Key Terms

Provide a one or two sentence answer for the following key terms.

RESISTANCE _____

_____

_____

AMBIVALENCE _____

_____

_____

IDENTIFYING RELATIONAL PATTERNS _____

_____

_____

VALIDATING RESISTANCE _____

_____

_____

TALKING ABOUT CONFLICTED FEELINGS _____

_____

_____

FORMULATING WORKING HYPOTHESES _____

_____

_____

EXTERNALIZING RESPONSIBILITY _____

_____

_____

APPROACHING INTERPERSONAL CONFLICT _____

_____

_____

DEFLECTION VS REFLECTION _____

_____

_____

"DIFFERENT" VIS A VIS CLIENT DIVERSITY _____

_____

_____

CONFLICT _____

_____

_____

THREE-STEP SEQUENCE FOR APPROACHING RESISTANCE _____

_____

_____

SOURCES OF RESISTANCE _____

_____

_____

RELATIONAL REENACTMENTS _____

_____

_____

CREATING INTERPERSONAL SAFETY _____

_____

_____

SHAME VS GUILT _____

_____

_____

SHAME-ANXIETY _____

_____

_____

SHAME-RAGE CYCLE _____

_____

_____

SURVIVOR GUILT _____

_____

_____

SEPARATION GUILT _____

_____

_____

Section B: Key Concepts

Consider each of the following key concepts. Be prepared
to write a two to three paragraph essay for each of the
following questions.

1.   Discuss the cardinal role of shame in clients' resistance
      and defense. How can therapists intervene with this
      sensitive feeling so that therapy can progress?

_____

_____

_____

_____

_____

_____

_____

_____

2.   Discuss reasons for therapists' reluctance to address
      resistance.

_____

_____

_____

_____

_____

_____

3.   Clients may miss appointments or come late because of reality-based constraints, personal conflicts over some aspect of entering treatment, or both. Discuss effective ways to intervene when <u>both</u> sources of "resistance" are operating.

_____

_____

_____

_____

_____

_____

_____

4.   Susan is a 32 year-old recently divorced mother of two. She is working two jobs and is seeing you at a reduced fee. You have been counseling her for seven weeks. She has canceled three of her seven appointments and arrived late for the remaining sessions. How would you begin to formulate "working hypotheses" for this client? Elucidate your rationale for these working hypotheses. Discuss effective and ineffective ways to address Susan's apparent resistance to treatment?

_____

_____

_____

_____

_____

_____

Section C: Sample Multiple Choice and True/False Questions

        Answer the following sample multiple choice and true/false questions.

1.   A client may be resisting treatment when:
      a.   he/she has difficulty participating.
      b.   he/she becomes confrontive with his/her family.
      c.   he/she displays a "moving toward" or pleasing attitude toward the therapist.
      d.   he/she avoids eye-contact during the first session.

2.   In order to honor a client's resistance, the therapist needs to:
      a.   reassure the client.
      b.   provide reasons for the resistance to the client.
      c.   reframe the resistance as having an adaptive response to a developmental conflict.
      d.   discuss the resistance logically with the client.

3.   Clients often interpret their own resistance as being shameful and want to avoid the topic.
      a.   true
      b.   false

        Answers: 1.a, 2.c, 3.a

PART II: SELF-REFLECTIONS

The material presented in the text is often personally evocative and challenging. Below are questions designed to help students process and integrate their reactions to the information.

1.   What is your initial response when someone is critical or disapproving of you or what you have done? What are the central feelings evoked in you? How do you tend to behaviorally respond to them?

_____

_____

_____

_____

_____

_____

_____

_____

_____

_____

_____

_____

_____

_____

2. List the personal or emotional reactions evoked in you when
    you have a problem you can't solve on your own and need to
    ask others for help. What do you tend to do, or how do you
    manage/express these reactions with others?

    _____

    _____

    _____

    _____

    _____

    _____

    _____

    _____

    _____

    _____

    _____

    _____

    _____

    _____

    _____

    _____

    _____

PART III: SUPPORTIVE READINGS AND CASE STUDIES

Shame is one of the least discussed but most important features of clients' presenting symptoms and problems. The following case study presents a client with shame-based dynamics. In this example, *the clinician and client together overtly identify or name the client's shame and remain emotionally connected throughout this process.* This illustrates how a therapist can provide clients with a reparative emotional experience that does not confirm the clients' old maladaptive templates or expectations of being rejected or abandoned.

Section A: Shame Dynamics: Case Study of Hank

<u>INTRODUCTION:</u> Hank, a 21 year-old, white college junior, was referred by a colleague who had worked with his family during his parents' divorce. The family therapist referred him to me to work with his long-standing anxiety and depression.

My most basic theoretical assumption is that maladaptive relational patterns and injuries to self-esteem are learned in interaction with significant others and can be unlearned in current interaction with others who are important. In particular, the interaction with the therapist, in which the therapist provides the client with a reparative experience that does not reenact the old relational patterns or confirm maladaptive templates, is the primary vehicle for enduring

change. Thus, the client (1) has a reparative experience that
does not repeat old expectations and/or patterns; (2) works
together with the therapist to generalize this experience to
other relationships by identifying and clarifying how they are
disrupting other relationships; and (3) is helped to formulate
and rehearse more adaptive responses and to anticipate
situations where old patterns are likely to be evoked.

This work was organized around several major life
themes/issues with specific interventions being helpful in each
area. I believe that addressing these major themes will be the
most effective way of conveying the essence of this work.
Obviously, these themes are overlapping and interwoven.

CLIENT DESCRIPTION: Hank was casually dressed in Levi's, Nikes,
and a tank top. He was 6' tall, bright, and attractive. My
primary impression of Hank was that he was extremely cautious
and fearful. He spoke in a hesitant and faltering manner, but
the complexity of his communication and his easy use of a well-
developed vocabulary suggested considerable intellectual
ability. Affect (particularly anxiety and depression) seemed
close to the surface although Hank was guarded in his
expression.

Hank came from a family with two siblings. His sister, who
is three years older, works out of the country. His brother,
who is three years younger, is a waiter. Hank's mother and

father were both 30 years old when he was born. His preschool years were spent living in the blue collar section of a mid-sized West Coast city. He stated that the neighborhood was a poor fit for him and that he had little in common with the other children or families in the community.

When he was 6, his family moved to a larger city and bought a home in a working-class section that had a Latino barrio on one side and a more affluent "yuppie" neighborhood on the other. Initially, Hank went to a primarily Latino school but was often taunted and bullied. His parents then sent him to a Catholic school. Although bullied less frequently, he was still an "outsider" and did not succeed in sports or with peers. He was painfully shy and socially awkward -- becoming vigilant about potential rejection and guarded about revealing himself. His parents were impatient with his insecurities, were emotionally unsupportive in general, and seemed to communicate that they did not want to have to "cater" to his "special" needs.

Early in high school "Dad divorced all of us." After leaving, he maintained superficial contact with his children for a while, then gradually faded away. Hank was deeply affected by this abandonment. Upon making the subsequent discovery that his father was bisexual, Hank became profoundly ashamed of him and, in turn, further ashamed of himself. Hank's anxiety and depression crystallized at this point, and he became

increasingly insecure about his own adequacy and identity
issues. Hank made it through high school and community college
with help from his high school band teacher who reached out to
him and affirmed his musical abilities and native intelligence.

In addition to academic ability, Hank had significant
musical ability. With the help of a scholarship and additional
money he had saved, he spent a summer at a reputable School of
Music in the North East. This was one of the best experiences
of his life, and he felt he learned an enormous amount about
music and about himself while he was away from home.

Hank's mother came from a prominent Southern family that
had in its history a Vice President, a Governor, and a Senator.
She felt that she had married beneath herself and resented the
lifestyle that her marriage brought with it. She was a woman in
her early 50's who had proven ability and occupied a responsible
managerial position in a large corporation. Apparently she had
considerable wit and charm and was described by Hank as a
"classy" lady. Her definition of herself and her children as
"Brahmans" had been two-edged -- a brittle source of pride that
brought with it extensive performance demands and fear of
inadequacy.

After the divorce she turned to her children, and Hank in
particular, for emotional gratification and comfort, which both
flattered and shackled him. His mother could at times be

genuinely warm and caring, and at other times, narcissistically self-absorbed and overtly rejecting. Hank experienced a maddening inconsistency as his mother alternately made him "special" and was then critical and rejecting. His sister disengaged from their mother and moved to another part of the country. From this distance the sister became able to emotionally support Hank in a consistent correspondence that was mutually valued. Hank's brother fought his way out of this web by acting out.

Hank indicated that most people who met his father would react to him as an "odd person." Father had ample supplies of charm and ability, but throughout his life he had difficulty getting along with others, which led to one self-destructive situation after another. Estranged from all members of his own family for years, he was a dreamer and a creative inventor who held several patents but was unable to turn any of them into financial gain.

Hank described his father as self-centered, arrogant, combative, destructively competitive with his sons, and shallow in his relationships. He had a marked tendency to externalize blame and could fly into a rage when his wishes were not honored. Hank's primary recollection was of being criticized by him frequently, being very frightened of him ("the primary emotion of my childhood"), and doing his best to stay out of his

way and not upset him. Just prior to entering treatment, Hank
had located his father -- seeking a rapprochement. Sadly, his
father was unable to sustain his initially positive response.
This disappointment exacerbated Hank's anxiety and depression
and led him to therapy.

The relationship between Hank's parents involved intense
power struggles. There were frequent arguments and occasional
physical fights. The children were pulled into the parental
battles and pressured to take sides.

TREATMENT: Hank's presenting complaints were anxiety and
depression. It quickly became clear to both of us that these
symptoms were related to his intensely ambivalent feelings about
his mother, coming to terms with the emotional deprivation he
had experienced with his father, and his own deep-seated
feelings of inadequacy and worthlessness. Treatment focused on
the following themes.

Establishing a working alliance: It quickly became evident
that several issues related to trust and safety had to be
addressed in our relationship. His relationship with his father
led me to hypothesize that there would be fears of attack,
competition, abandonment, and control. His relationship with
his mother led me to hypothesize that he expected he could only
be connected to me if he met my needs and did not have his own
voice. He was hypervigilant for criticism from me and often

misperceived my reflective, empathic attempts as critical or judgmental.

Given Hank's templates, therapy was structured in a way to give him maximum control and foster his initiative. To meet these treatment goals I indicated that: (1) I knew that some things were difficult to talk about and that he should set the pace; (2) I did not wish to be intrusive and if he asked me to back off I would do so immediately; (3) I would appreciate feedback if I were misunderstanding or misinterpreting what he was telling me; and (4) I was interested in knowing what he considered the most important topics to discuss. I repeatedly communicated that our work together was a collaborative effort. I was aware that whenever Hank became anxious in our relationship he misperceived me as being controlling or critical and was again shamed by this. Over a period of several months (and after testing me repeatedly to see if I meant what I said), an effective therapeutic alliance was developed with Hank taking an active lead in exploring his issues.

Relationship with mother -- integrating ambivalence: I sought opportunities to assure him that: (1) my motives for examining his relationship with his mother were not the same as his father's motives (i.e., taking sides or "mother bashing"); (2) I understood the importance of that relationship to him and recognized its positive aspects; (3) I would prefer to see the

relationship change in healthy ways rather than be lost; and (4)
however he decided to manage the relationship would be his
decision. As these assurances gradually took hold, Hank was
able to begin examining this relationship.

Hank genuinely loved his mother but was deeply hurt by her.
On the one hand, he was ashamed of his intense need of her and
how easily she could disengage from him. At the same time, he
was very protective of her and, to preserve the tenuous ties,
was unwilling to acknowledge conflict with her. Hank was afraid
that if he advocated on his own behalf and pursued his own
interests he would be "betraying" his mother, abandoning her
like his father had done, and would then be deserving of the
critical rejection that would likely follow.

Hank had few experiences of personal efficacy in his life;
he had been demeaned by his father, humiliated by his peers, and
inconsistently valued by his mother. These all conspired to
create a shame-based sense of self, which was defended against
by the symptoms of anxiety and depression. Working with Hank to
acknowledge and accept the reality of both the positive and
negative aspects of his mother permitted him to more
realistically address his anxiety about having his pervasive
sense of shame exposed to others. My ability to remain caring
and affirming of Hank, especially during these most sensitive
moments in treatment when I was honored to be able to bear

witness to his shame, permitted him to become more accepting and compassionate toward himself. *Having revealed his shame and finding compassion rather than the contempt he had expected, Hank's anxiety abated.*

With my encouragement, Hank took the initiative to talk with his mother about their relationship, something that had never occurred before. Fortunately, rather than just being threatened by his directness she was able to engage in some self-examination, and they began changing the way they related to each other.

Relationship with father -- deprivation, anger, grief, and disappointment: Hank's father had been emotionally abusive toward him. As this emerged in treatment, I responded with validation and compassion, which initially evoked shame and rage in Hank. As we incrementally worked through these feelings together, however, the profound sadness and deprivation emerged, and we entered a period of grief and mourning. As Hank was able to acknowledge how much he had been hurt and how much he had lost, his life-long depression diminished. It was with great pleasure that I observed increasing optimism and initiative.

Self-efficacy and shame issues: As I got to know Hank better, I was struck by how much talent and ability he had and how little he owned it. His experiences with a denigrating father who called him "stupid" and the humiliating experiences

of being bullied and ostracized at school furthered his sense of
self as worthless and inadequate. His father needed to be
center of the stage and "top dog." In order to maintain some
tie, ward off his father's anger, and not "outdo" his father,
Hank minimized his musical ability, competency in computers and
math, appetite for books, and "hobbies" like taking foreign
language courses.

During one of our sessions, I made an almost offhand
comment, "Hank, don't you know that you have an unusually wide
range of talents and abilities?" This statement made an
enduring impact on him. He reported later being "stunned" by
that comment, and we spent a great deal of time discussing why
he "kept his foot on the brake rather than the accelerator when
he had so much horsepower."

In sum, my input involved (1) helping him make sense of his
underachievement; (2) helping him understand why he had been
anxious and depressed throughout his life; and (3) supporting
his competency and encouraging expression of his own voice. I
let him know that our respective competencies were not "yoked"
together and that I valued his adequacy instead of needing to
compete with it. I let him see the pleasure it gave me when he
took initiative and expressed his own mind. We also talked
about the need to take modulated risks and better assess his

current relationships as a way of determining if his fears were based in current reality or past history.

TERMINATION: The courage and energy with which Hank responded to our reparative relationship was impressive. He began to force himself to talk in groups and to initiate new friendships. He applied himself diligently to his studies for the first time and obtained a 4.00 GPA two terms in a row. Musically, he tried out for a local talent show and took third place in it.

With these treatment goals well under way, we entered the termination phase. This was particularly important in light of Hank's abandonment experience with his father and inconsistent tie with his mother. During the last part of therapy, we met on an every other week basis so that he could achieve some independence from therapy as well as maintain a connection to me. I left it up to Hank to choose the termination date, and we talked at some length about what our work together had meant to both of us. The criterion that I suggested for termination was that we end when Hank felt internally more secure and trusted that a constructive internal growth process was continuing. A related criterion in my own mind was that he be developmentally in step with his peers. I believe that by our last session both of these criteria had been met.

If Hank were your client, what feelings would be evoked in you by his shame? What feelings could be evoked in you by naming and addressing his shame-based sense of self?

_____

_____

_____

_____

_____

_____

_____

_____

_____

_____

_____

_____

_____

_____

_____

_____

Section B: Sample of Nonshaming Contact Letter to Resistant
             Clients

Current Date

Client's Name
1234 Counseling Way
Therapy Village, CA 92000

Dear (Client's first name),

     You have not attended our last (#) sessions on (fill in
date) and have not called to cancel or returned my calls. I
have missed meeting with you and am wondering if you have
decided not to continue in treatment at this time. I would like
for you to call me at (phone #) before (date) so we can schedule
1 more session -- either to say good-by to each other or to
discuss what we could do differently together to make therapy
more helpful for you. If I haven't heard from you by (date), I
will assume that you have decided not to continue treatment at
this time and I will not hold our appointment time any longer.
I hope I hear from you, but if not, I have enjoyed meeting with
you and wish you the very best in the future.

                    Yours truly,

                    Counselor's Name

Section C: Suggested Readings on Shame

Balcom, D., Lee, R.G., & Tager, J. (1995). The systematic
     treatment of shame in couples. *Journal of Marital and
     Family Therapy, 21, 1,* 55-65.
Shame cycles in couples are discussed.

Dutton, D.G. (1995). *The batterer: A psychological profile* (pp.
     78-93). San Francisco: Basic Books.
**This reading highlights the all-important shame/rage cycle, and
the role of shame in domestic violence.**

Fowler, J.W. (1996). *Faithful change* (pp.113-121). Tennessee:
     Abingdon Press.
**This reference elucidates perfectionistic defenses against
shame.**

Jordan, J. (Ed.) (1997). *Women's growth in diversity: More
     writings from the Stone Center* (pp. 138-161). New York: The
     Guilford Press.
**This reading articulates the cardinal role of exposure in shame
and illuminates women's subjective experience of shame.**

Karen, R. (1992, February). Shame. *The Atlantic Monthly,* pp.
     40-70.
**This reading provides an informative introduction/overview of
shame and is wellwritten in lay terms.**

PART IV: JOURNAL ENTRY

Use the following section to further integrate the material presented, to address unresolved questions or issues, and to explore further your personal reactions.

_____

_____

_____

_____

_____

_____

_____

_____

_____

_____

_____

_____

_____

_____

CHAPTER FOUR: AN INTERNAL FOCUS FOR CHANGE

PART I: STUDY GUIDE

Part I will help students prepare for course examinations. It consists of three sections: a list of key terms to define; essay questions for central concepts emphasized in the text; and sample multiple choice and true/false questions.

Section A: Key Terms

Provide a one or two sentence answer for the following key terms.

INTERNAL VS EXTERNAL FOCUS _____

_____

_____

SHAME-PRONE _____

_____

_____

SELF-EFFICACY _____

_____

_____

VALIDATION _____

_____

_____

REENACTMENTS/RECAPITULATION _____

_____

_____

CLIENTS' INITIATIVE _____

_____

_____

PROCESS DIMENSION _____

_____

_____

HIERARCHICAL RELATIONSHIP VS SHARED CONTROL _____

_____

_____

DEPENDENCY-FOSTERING VS INDEPENDENCE-FOSTERING _____

_____

_____

TOLERATING AMBIGUITY _____

_____

_____

CORRECTIVE EMOTIONAL EXPERIENCE _____

_____

_____

TRACKING CLIENT ANXIETY _____

_____

_____

PRECIPITATING EVENTS _____

_____

_____

Section B: Key Concepts

     Consider each of the following key concepts. Be prepared
to write a two or three paragraph essay for each of the
following questions.

1.   Discuss several reasons why clients may externalize their
     problems. Suggest different ways of responding to this
     behavior that may help clients move toward a more internal
     focus.

_____

_____

_____

_____

_____

_____

_____

2.   Suggest several reasons why counselors might collude
     with their clients in externalizing their problems. How
     might this recapitulate clients' maladaptive relational
     patterns?

_____

_____

_____

_____

_____

_____

_____

3. Discuss the potential problems or reenactments that are
    likely to occur with a "hierarchical" therapeutic
    relationship. How does a collaborative or working alliance
    foster clients' initiative and self-efficacy?

    _____

    _____

    _____

    _____

    _____

    _____

    _____

    _____

4. How do clients tend to express their anxiety during
    therapy? Discuss effective methods for tracking and
    approaching this anxiety with clients.

    _____

    _____

    _____

    _____

    _____

    _____

    _____

    _____

Section C: Sample Multiple Choice and True/False Questions

    Answer the following sample multiple choice and true/false questions.

1.   In working with a client who is externalizing his/her problems, a therapist must first:
    a.   help the client take an internal focus.
    b.   acknowledge the complaints as valid concerns.
    c.   give the client several solution options.
    d.   find the repetitive relational conflict involved in the client's interactions.

2.   When therapists track or actively attend to their clients' anxiety, they are better able to:
    a.   determine an appropriate diagnosis for their clients.
    b.   establish a relationship that meets their clients' perceived needs.
    c.   assess and identify clients' central conflicts.
    d.   identify the person responsible for the clients' problems.

3.   A hierarchical helper-helpee mode of therapy usually helps to alleviate clients' concerns.
    a.   true
    b.   false

        Answers: 1.b, 2.c, 3.b

PART II: SELF-REFLECTIONS

The material presented in the text is often personally evocative and challenging. The questions below are designed to help students process and integrate their reactions to this information.

1.   Think of the most important success experience from your childhood or adolescence. Recall how your mother and your father (or other central caregivers) responded to this important event in your life. Be as specific as you can about what significant others said or did, both overtly and covertly, to affirm or invalidate your happiness and success.

_____

_____

_____

_____

_____

_____

_____

_____

_____

_____

_____

_____

1.   Think of what it means for you to have a sense of self-efficacy or personal mastery. Thinking in terms of two time periods, both as a child and now as an adult, create two lists. For each time period, create a list of: (1) the interpersonal interactions that have made you feel effective and in charge of yourself and (2) those that have left you feeling disempowered. Both as a child and as an adult, what kinds of things did you tend to say and do following each interaction?

_____

_____

_____

_____

_____

_____

_____

_____

_____

_____

_____

_____

_____

_____

_____

_____

3.   Recall a time when you failed to meet or live up to the
     expectations of an important other. How did this make you
     feel, and what did you do to cope with these feelings? In
     looking back, what could you and others have done to help
     you in this situation?

_____

_____

_____

_____

_____

_____

_____

_____

_____

_____

_____

_____

_____

_____

_____

_____

_____

PART III: SUPPORTIVE READINGS AND CASE STUDIES

Therapists need to develop certain interpersonal process skills to work effectively with clients. The following questionnaire delineates these important interpersonal process skills. Use this assessment to systematically evaluate your current level of skill in working along each of these dimensions. Students may wish to videotape a session with a client or role-play with a partner and assess their interaction.

Section A: Clarifying Therapist's Interpersonal Style

1.   Therapist makes process comments or finds other ways to openly discuss distortions, misunderstandings, and other potential problems that may be occurring between the client and the therapist.

1	2	3	4	5	6	7
not at all characteristic			characteristic			extremely characteristic

2.   Therapist actively tries to understand the client's socio-cultural context and how race, religion, and gender have shaped *her** subjective worldview.

1	2	3	4	5	6	7
not at all characteristic			characteristic			extremely characteristic

3.   Therapist accurately identifies and reflects the central meaning or emotional message in what the client has just relayed.

1	2	3	4	5	6	7
not at all characteristic			characteristic			extremely characteristic

4.   Therapist helps the client relay her narrative and express thoughts and feelings.

1	2	3	4	5	6	7
not at all characteristic			characteristic			extremely characteristic

Throughout the questionnaire the female pronoun is used to refer to the client.

5.   Therapist helps the client explore and discuss her personal
      reactions toward the therapist.
         1   2   3   4   5   6   7
      not at all   characteristic   extremely
      characteristic   characteristic

6.   Therapist has difficulty attending to how the client may be
      interacting with the therapist in the same problematic ways
      that she describes doing with others.
         1   2   3   4   5   6   7
      not at all   characteristic   extremely
      characteristic   characteristic

7.   Therapist has difficulty following the client's lead and
      staying close to the problems and issues that the client
      reports as relevant or significant in her life right now.
         1   2   3   4   5   6   7
      not at all   characteristic   extremely
      characteristic   characteristic

8.   Therapist explores developmental events as they arise in
      the conversation naturally, rather than leading the client
      back to historical connections.
         1   2   3   4   5   6   7
      not at all   characteristic   extremely
      characteristic   characteristic

9.   Therapist helps the client focus inward on her own thoughts
      and feelings.
         1   2   3   4   5   6   7
      not at all   characteristic   extremely
      characteristic   characteristic

10.   Therapist has difficulty inviting the client to express
      whatever feelings she may be experiencing as they occur in
      the session.
         1   2   3   4   5   6   7
      not at all   characteristic   extremely
      characteristic   characteristic

11.   Therapist encourages the client to explore feelings and
      thoughts about their current interaction and what is
      happening in the therapeutic relationship.
         1   2   3   4   5   6   7
      not at all   characteristic   extremely
      characteristic   characteristic

12. Therapist is able to extend herself and actively reach out when necessary to maintain the client's engagement in a collaborative relationship.

      1   2   3   4   5   6   7
    not at all   characteristic   extremely
    characteristic   characteristic

13. Therapist is nonjudgmental and responds to the client in an accepting and understanding manner.

      1   2   3   4   5   6   7
    not at all   characteristic   extremely
    characteristic   characteristic

14. Therapist attempts to conceptualize a treatment focus by formulating a working hypothesis about the maladaptive relational patterns and interpersonal themes that reoccur in the client's life.

      1   2   3   4   5   6   7
    not at all   characteristic   extremely
    characteristic   characteristic

15. Therapist identifies and explores problematic relational patterns that might constitute a generic conflict or organizing theme in the client's interpersonal relationships.

      1   2   3   4   5   6   7
    not at all   characteristic   extremely
    characteristic   characteristic

16. Therapist attempts to link recurrent patterns of behavior between the client and others to the interaction between the client and therapist.

      1   2   3   4   5   6   7
    not at all   characteristic   extremely
    characteristic   characteristic

17. Therapist looks for unifying themes or patterns that may link events that initially appear to be unrelated.

      1   2   3   4   5   6   7
    not at all   characteristic   extremely
    characteristic   characteristic

18. When the client appears to become defensive or resistant, therapist helps the client explore what the danger or threat is that may have just been evoked.

          1   2   3   4   5   6   7
     not at all   characteristic   extremely
     characteristic   characteristic

19. Therapist is reluctant to focus the client away from complaining about or describing the problematic behavior of others and toward the client's own personal reactions.

          1   2   3   4   5   6   7
     not at all   characteristic   extremely
     characteristic   characteristic

20. Therapist responds to the client's global descriptions or generalized statements about themselves and others by seeking further specificity or concrete illustrations.

          1   2   3   4   5   6   7
     not at all   characteristic   extremely
     characteristic   characteristic

21. Therapist is empathic and tries to understand the personal or unique meanings of the client's experience from the client's subjective worldview.

          1   2   3   4   5   6   7
     not at all   characteristic   extremely
     characteristic   characteristic

22. Therapist is emotionally available and conveys "presence" as the client relays her narratives.

          1   2   3   4   5   6   7
     not at all   characteristic   extremely
     characteristic   characteristic

23. Therapist cannot be a "participant/observer" who is simultaneously empathic and objective.

          1   2   3   4   5   6   7
     not at all   characteristic   extremely
     characteristic   characteristic

24. Therapist invites the client to discuss what she is thinking about the therapist, what she is thinking about something the therapist has done, or what she is thinking the therapist might be feeling.

          1   2   3   4   5   6   7
     not at all   characteristic   extremely
     characteristic   characteristic

25. Therapist creates <u>immediacy</u> by using self-involving statements or sharing her own reactions to the client.

      1   2   3   4   5   6   7

not at all   characteristic   extremely

characteristic   characteristic

26. Therapist encourages the client to be an active, equal partner in understanding problems and initiating changes.

      1   2   3   4   5   6   7

not at all   characteristic   extremely

characteristic   characteristic

27. Therapist helps the client discern discrepancies between her public persona or her social roles <u>and</u> her own authentic voice and genuine experience.

      1   2   3   4   5   6   7

not at all   characteristic   extremely

characteristic   characteristic

28. Therapist considers unwanted ways that significant others have responded to the client in the past and uses this awareness to provide a new or reparative response to the client.

      1   2   3   4   5   6   7

not at all   characteristic   extremely

characteristic   characteristic

29. Therapist has difficulty formulating working hypotheses about how the client's relational patterns could interact with the therapist's own personal issues to impede treatment.

      1   2   3   4   5   6   7

not at all   characteristic   extremely

characteristic   characteristic

30. Therapist helps the client appreciate how her defensive style was originally necessary and adaptive but is no longer needed in many current relationships.

      1   2   3   4   5   6   7

not at all   characteristic   extremely

characteristic   characteristic

31. Therapist demonstrates the cognitive flexibility and wide
     emotional range necessary to respond to the client's
     varying needs.

          1   2   3   4   5   6   7
     not at all   characteristic   extremely
     characteristic   characteristic

32. Therapist evaluates the effectiveness of her interventions
     by systematically evaluating the client's reactions to
     them.

          1   2   3   4   5   6   7
     not at all   characteristic   extremely
     characteristic   characteristic

33. Therapist feels she can be authentic with the client
     without feeling distanced or ingenuine by the constraints
     of the therapeutic role.

          1   2   3   4   5   6   7
     not at all   characteristic   extremely
     characteristic   characteristic

34. Therapist is unable to balance the two-sided challenge of
     being forthright and direct with the client while remaining
     empathic and respectful.

          1   2   3   4   5   6   7
     not at all   characteristic   extremely
     characteristic   characteristic

PART IV: JOURNAL ENTRY

Use the following section to further integrate the material presented, to address unresolved questions or issues, and to further explore your personal reactions.

_____

_____

_____

_____

_____

_____

_____

_____

_____

_____

_____

_____

_____

CHAPTER FIVE: RESPONDING TO CONFLICTED EMOTIONS

PART I: STUDY GUIDE

Part I will help students prepare for course examinations. It consists of three sections: a list of key terms to define; essay questions for central concepts; and sample multiple choice and true/false questions.

Section A: Key Terms

Provide a one or two sentence answer for the following key terms.

DIRECTLY ADDRESSING FEELINGS _____

_____

_____

UNDERLYING AFFECT _____

_____

_____

NONVERBAL CUES _____

_____

_____

OPEN-ENDED QUESTIONS _____

_____

_____

CLOSED-ENDED QUESTIONS _____

_____

_____

SUBJECTIVE WORLDVIEW _____

_____

_____

UNDIFFERENTIATED FEELING STATES _____

_____

_____

CLARIFYING FEELINGS _____

_____

_____

MIRRORING _____

_____

_____

EMPATHIC ATTUNEMENT _____

_____

_____

INCONGRUENCE _____

_____

_____

PREDOMINANT/CENTRAL AFFECT _____

_____

_____

OLD WOUND _____

_____

_____

COMPACTED PHRASE _____

_____

_____

RECURRENT AFFECTIVE THEMES _____

_____

_____

CHARACTEROLOGICAL AFFECT _____

_____

_____

AFFECTIVE SEQUENCE OR CONSTELLATION _____

_____

_____

REACTIVE ANGER _____

_____

_____

SHAME CYCLE _____

_____

_____

INTERNALLY RESOLVING CONFLICT _____

_____

_____

OBSERVING SELF _____

_____

_____

HOLDING ENVIRONMENT _____

_____

_____

EXPLORING THE RELATIONAL THREAT _____

_____

_____

CLARIFYING CLIENT DEFENSES _____

_____

_____

ASSUMING RESPONSIBILITY FOR CLIENTS' PAIN _____

_____

_____

NONVERBAL EMOTIONAL CONNECTEDNESS _____

_____

_____

EMOTIONAL PRESENCE _____

_____

_____

CORRECTIVE EMOTIONAL EXPERIENCE _____

_____

_____

DISCONFIRMING PATHOGENIC BELIEFS _____

_____

_____

CHANGE FROM THE INSIDE OUT _____

_____

_____

MALADAPTIVE RELATIONAL PATTERNS _____

_____

_____

FAMILIAL ROLES AND RULES _____

_____

_____

RECOGNIZING COUNTERTRANSFERENCE PROPENSITIES _____

_____

_____

Section B: Key Concepts

Consider each of the following key concepts. Be prepared to write a two to three paragraph essay for each of the following questions.

1.   Many clients try to protect their parent(s) from realizing the negative or hurtful impact the caregiver has had on them during childhood. How could this guilt and family loyalty prevent clients from experiencing their own conflicts or painful feelings in therapy? Suggest effective and ineffective ways of intervening with these clients.

_____

_____

_____

_____

_____

_____

_____

_____

_____

_____

_____

_____

_____

_____

2.   Discuss the concept of affective constellations.
     Illustrate one potential triad of interrelated feelings,
     and describe how this pattern or sequence of feelings might
     present in treatment.

_____

_____

_____

_____

_____

_____

_____

_____

_____

3.   Describe what is meant by a "holding environment." How can
     a therapist accomplish this with a client, and why is it
     important to the client's treatment?

_____

_____

_____

_____

_____

_____

_____

_____

4.   Discuss factors that lead to countertransference reactions
     in therapy. Illustrate effective and ineffective ways
     therapists can manage these reactions.

_____

_____

_____

_____

_____

_____

_____

_____

_____

_____

5.   Suggest effective and ineffective ways of responding to
     clients who frequently talk about their feelings without
     actually experiencing them.

_____

_____

_____

_____

_____

_____

_____

Section C: Sample Multiple Choice and True/False Questions

Answer the following sample multiple choice and true/false questions.

1.   The most effective way for therapists to approach their client's affect is to:
     a.   respond by seeking more information about the content of what the client just said.
     b.   respond to the feeling that the client is currently experiencing.
     c.   respond by telling the client he or she will feel differently soon.
     d.   respond by giving the client concrete ways to change the affect.

2.   When therapists find they are repeatedly having difficulty with countertransference and supervision does not help, they should:
     a.   encourage the client to discuss other topics.
     b.   ignore their countertransference.
     c.   seek therapy themselves.
     d.   talk about their countertransference with their spouse.

3.   When clients risk exposing their pain, vulnerability, or shame, and the therapist responds with kindness and understanding, clients become too dependent on the therapist.
     a.   true
     b.   false

Answers: 1.b, 2.c, 3.b

PART II: SELF-REFLECTIONS

The material presented in the text is often personally
evocative and challenging. The questions below are designed to
help students process and integrate their reactions to this
information.

1.   Identify the most shameful experience of your childhood
      that you can recall. In retrospect, what could concerned
      others have said or done to help you recover from this
      shame reaction?

_____

_____

_____

_____

_____

_____

_____

_____

_____

_____

_____

_____

_____

_____

2. Therapists often avoid client's feelings because they: (1) experience guilt about having the client reveal their very painful affects, and/or (2) believe they are inadequate to respond to them because they think they have to "fix" them. Try to identify developmental experiences in your family-of-origin in which you were made to feel responsible for the pain of other family members or were expected to "fix" or take away their pain.

_____

_____

_____

_____

_____

_____

_____

_____

_____

_____

_____

_____

_____

_____

_____

_____

_____

3.   Anger is often a reactive response to pain, helplessness, frustration, or other feelings. Explore how these other primary affective states may be masked by your reactive anger.

_____

_____

_____

_____

_____

_____

_____

_____

_____

_____

_____

_____

_____

_____

_____

_____

_____

_____

4.   Think of a recurrent or significant event or incident from
     your childhood that caused you emotional pain. How did
     each of your parents (or other primary caregivers) respond
     to you? How did your behavior, thoughts, feelings, or
     actions change as a result of their response?

_____

_____

_____

_____

_____

_____

_____

_____

_____

_____

_____

_____

_____

_____

_____

_____

_____

5.   Think about the rules governing emotional expression in
      your family-of-origin. What feelings can be expressed, and
      which are unacceptable? Of the "acceptable" feelings,
      when, how, and to whom can they be expressed? If you were
      to violate unspoken familial rules and express one of the
      "unacceptable" feelings, how do you expect others to
      respond to you?

_____

_____

_____

_____

_____

_____

_____

_____

_____

_____

_____

_____

_____

_____

_____

_____

6. Recalling your family background, what was the most
    threatening or taboo affect you can recall? When clients
    express this feeling toward you, what are you likely to
    feel inside, and how are you likely to respond?

_____

_____

_____

_____

_____

_____

_____

_____

_____

_____

_____

_____

_____

_____

_____

_____

_____

_____

PART III: SUPPORTIVE READINGS AND CASE STUDIES

This exercise is designed to help you anticipate and manage your own reactions to evocative material or affect that a client might present. Use each of the following statements to assess your potential countertransference propensities.

Section A: Sensitivity to Countertransference Propensities

1.   If a client becomes critical or angry toward me, it is
      often difficult for me to remain open and nondefensive.
      1   2   3   4   5   6   7
      Not At All   Extremely
      Characteristic   Characteristic   Characteristic

2.   When my clients leave our session in pain, I often feel as
      though I have not done enough to help them.
      1   2   3   4   5   6   7
      Not At All   Extremely
      Characteristic   Characteristic   Characteristic

3.   In overt or covert ways, I tend to move clients away from
      experiencing or expressing their painful feelings.
      1   2   3   4   5   6   7
      Not At All   Extremely
      Characteristic   Characteristic   Characteristic

4.   I try to encourage my clients to express negative feelings
      (e.g., anger, disapproval) they may be having toward me.
      1   2   3   4   5   6   7
      Not At All   Extremely
      Characteristic   Characteristic   Characteristic

5.   I am often reluctant to speak forthrightly and address what
      is going on between the client and me, even though the
      client would benefit from such directness.
      1   2   3   4   5   6   7
      Not At All   Extremely
      Characteristic   Characteristic   Characteristic

6.   I do not give in to clients' wants or demands; instead I
      tend to set appropriate limits and boundaries.
      1   2   3   4   5   6   7
      Not At All   Extremely
      Characteristic   Characteristic   Characteristic

7.   I tend to avoid caring deeply or making emotional contact
     with clients because it might lead to a loss of appropriate
     boundaries or foster dependency.
     1   2   3   4   5   6   7
     Not At All   Extremely
     Characteristic   Characteristic   Characteristic

8.   I tend to find myself in power struggles or subtle control
     battles with my clients.
     1   2   3   4   5   6   7
     Not At All   Extremely
     Characteristic   Characteristic   Characteristic

9.   I am uncomfortable with ambiguity and tend to direct what
     is going to happen next in my counseling interactions.
     1   2   3   4   5   6   7
     Not At All   Extremely
     Characteristic   Characteristic   Characteristic

10. It is often hard for me to move my clients toward an
     internal focus where they can explore their own
     contributions to their problem, even though I know it would
     be therapeutic.
     1   2   3   4   5   6   7
     Not At All   Extremely
     Characteristic   Characteristic   Characteristic

11. I tend to either talk too much or too little when I feel
     overly affected by my client's issues.
     1   2   3   4   5   6   7
     Not At All   Extremely
     Characteristic   Characteristic   Characteristic

12. I readily identify with my clients and have difficulty
     letting go of their problems while outside of the
     therapeutic setting.
     1   2   3   4   5   6   7
     Not At All   Extremely
     Characteristic   Characteristic   Characteristic

13. I am comfortable making self-involving statements or
     process comments about what is occurring between the client
     and me when I think it would be helpful.
     1   2   3   4   5   6   7
     Not At All   Extremely
     Characteristic   Characteristic   Characteristic

14. I readily identify with my clients and have difficulty
     letting go of their problems outside of the therapeutic
     setting.
     1   2   3   4   5   6   7
     Not At All   Extremely
     Characteristic   Characteristic   Characteristic

15. I have difficulty accepting clients' genuine expressions of
     warmth toward me.
     1   2   3   4   5   6   7
     Not At All   Extremely
     Characteristic   Characteristic   Characteristic

16. I often carry out the familial role of caretaker, rescuer,
     or peacemaker in my professional work as a therapist.
     1   2   3   4   5   6   7
     Not At All   Extremely
     Characteristic   Characteristic   Characteristic

17. I tend to minimize interpersonal conflicts that arise
     between my clients and me.
     1   2   3   4   5   6   7
     Not At All   Extremely
     Characteristic   Characteristic   Characteristic

18. It is often hard for me to acknowledge "mistakes" I have
     made with clients or times when I have misunderstood them.
     1   2   3   4   5   6   7
     Not At All   Extremely
     Characteristic   Characteristic   Characteristic

19. I tend to be patient rather than frustrated or irritated
     with clients who appear to be needy, dependent, or
     helpless.
     1   2   3   4   5   6   7
     Not At All   Extremely
     Characteristic   Characteristic   Characteristic

20. I tend to be quiet during sessions in order to avoid making
     mistakes.
     1   2   3   4   5   6   7
     Not At All   Extremely
     Characteristic   Characteristic   Characteristic

21. I usually remain accepting and engaged with clients who
     make choices that disagree with my values.
     1   2   3   4   5   6   7
     Not At All   Extremely
     Characteristic   Characteristic   Characteristic

22. I tend to become impatient or irritated when clients are
     slow at making change.
     1   2   3   4   5   6   7
     Not At All   Extremely
     Characteristic   Characteristic   Characteristic

23. I am often concerned that my clients will perceive me as
     unskilled or inadequate.
     1   2   3   4   5   6   7
     Not At All   Extremely
     Characteristic   Characteristic   Characteristic

24. Rather than tolerating ambiguity and seeing what emerges
     from the client, I often find myself filling silences and
     easing awkward moments out of my own discomfort.
     1   2   3   4   5   6   7
     Not At All   Extremely
     Characteristic   Characteristic   Characteristic

25. I tend to have trouble distilling my clients' core
     affective messages or reflecting the deeper meaning in
     their narratives.
     1   2   3   4   5   6   7
     Not At All   Extremely
     Characteristic   Characteristic   Characteristic

26. It tends to be hard for me to express my disagreements with
     clients.
     1   2   3   4   5   6   7
     Not At All   Extremely
     Characteristic   Characteristic   Characteristic

27. It is usually too uncomfortable for me to give clients
     unwanted but honest interpersonal feedback.
     1   2   3   4   5   6   7
     Not At All   Extremely
     Characteristic   Characteristic   Characteristic

28. When clients criticize or disapprove of me, I tend to justify or defend myself rather than fully listen to and explore their concerns.

    1   2   3   4   5   6   7
    Not At All   Extremely
    Characteristic   Characteristic   Characteristic

29. I tend to overqualify my comments or dilute the emotional impact of my interventions to avoid having too much influence or impact on my clients' lives.

    1   2   3   4   5   6   7
    Not At All   Extremely
    Characteristic   Characteristic   Characteristic

30. Even though clients initially may feel hurt, angry, or abandoned, I usually bring up the issue of our impending termination.

    1   2   3   4   5   6   7
    Not At All   Extremely
    Characteristic   Characteristic   Characteristic

31. I often think about ways in which my own issues could impact the therapeutic relationship.

    1   2   3   4   5   6   7
    Not At All   Extremely
    Characteristic   Characteristic   Characteristic

32. Rate the degree to which you believe you are aware of your countertransference propensities.

    1   2   3   4   5   6   7
    Not At All   Extremely
    Characteristic   Characteristic   Characteristic

33. How frequently do you believe you behaviorally act out your countertransference propensities.

    1   2   3   4   5   6   7
    Not At All   Extremely
    Characteristic   Characteristic   Characteristic

After rating yourself on each of these, return to those items for which you have rated yourself as highly characteristic. Reflect on the questions you found relevant and consider: (1) how you might express yourself or "present" with regard to these issues in an actual therapy setting with clients and (2) how each of these selected issues could limit your ability to respond effectively to your clients' emotions.

_____

_____

_____

_____

_____

_____

_____

_____

_____

_____

_____

_____

_____

_____

_____

PART IV: JOURNAL ENTRY

Use the following section to further integrate the material presented, to address unresolved questions or issues, and to further explore your personal reactions.

_____

_____

_____

_____

_____

_____

_____

_____

_____

_____

_____

_____

CHAPTER SIX: FAMILIAL AND DEVELOPMENTAL FACTORS

PART I: STUDY GUIDE

    Part I will help students prepare for course examinations. It consists of three sections: a list of key terms to define; essay questions for central concepts; and sample multiple choice and true/false questions.

Section A: Key Terms

    Provide a one or two sentence answer for the following key terms.

PARENTAL COALITION _____

_____

_____

INDIVIDUATION _____

_____

_____

CROSS-GENERATIONAL ALLIANCE _____

_____

_____

SHIFTING LOYALTY TIES _____

_____

_____

INTERGENERATIONAL BOUNDARIES _____

_____

_____

EMANCIPATION CONFLICTS _____

_____

SEPARATION GUILT _____

_____

_____

PARENTIFICATION _____

_____

_____

SEPARATENESS-RELATEDNESS DIALECTIC _____

_____

_____

PROJECTIVE IDENTIFICATION _____

_____

_____

AUTHORITATIVE PARENTING STYLE _____

_____

_____

PERMISSIVE PARENTING STYLE _____

_____

_____

AUTHORITARIAN PARENTING STYLE _____

_____

_____

INTERNALIZED PARENT _____

_____

_____

SHAME-BASED SENSE OF SELF _____

_____

_____

DISSOCIATED EGO STATE _____

_____

_____

LOVE WITHDRAWAL _____

_____

_____

INTERPERSONAL COPING STRATEGIES _____

_____

_____

PARENT BASHING OR BLAME _____

_____

_____

INTEGRATING AMBIVALENT FEELINGS _____

_____

_____

EMOTIONAL PRESENCE _____

_____

_____

MALADAPTIVE RELATIONAL PATTERNS _____

_____

_____

PATHOGENIC BELIEFS _____

_____

_____

REENACTMENT OF CLIENT'S CONFLICT _____

_____

_____

FAMILY-OF-ORIGIN WORK _____

_____

_____

GENOGRAM _____

_____

_____

Section B: Key Concepts

Consider each of the following key concepts. Be prepared to write a two or three paragraph essay for each of the following questions.

1. Discuss how cross-generational alliances and a lack of clear intergenerational boundaries may shape a client's presenting problems. How would you establish a treatment plan for such a client?

_____

_____

_____

_____

_____

_____

_____

2. Recall the three styles of parenting outlined in the chapter and consider potential consequences of each style for adults' development. How might each present in therapy?

_____

_____

_____

_____

_____

_____

3. Discuss implications for therapy when working with
    clients who have experienced pervasive and severe love
    withdrawal.

    _____

    _____

    _____

    _____

    _____

    _____

    _____

    _____

4. Discuss the dynamics of parentification. How might
    parentified adult clients behave in their current
    interpersonal relationships? How might this play out in
    the therapeutic relationship, and how could this affect
    your treatment plans?

    _____

    _____

    _____

    _____

    _____

    _____

    _____

5.   Discuss what is meant by the "separateness-relatedness dialectic" and how this concept could be useful to clinicians.

_____

_____

_____

_____

_____

_____

_____

_____

_____

_____

_____

_____

_____

_____

_____

_____

_____

_____

Section C: Sample Multiple Choice and True/False Questions

Answer the following sample multiple choice and true/false questions.

1.   When children are allowed to play one parent against the other, they:
      a.   learn that rules do not apply to them.
      b.   learn that rules always apply to them.
      c.   are rarely viewed as manipulative by others.
      d.   tend to become extremely high achievers.

2.   Clients who have been parentified as children may:
      a.   avoid any situations requiring them to act responsibly.
      b.   grow up to become care-giving professionals who are at risk of early professional "burnout."
      c.   relinquish control easily and are comfortable in mutually reciprocal relationships.
      d.   be inattentive and unresponsive to the therapist.

3.   The principal reason children are scripted into delimiting familial roles and the reason family myths are established is to help parents defend against their own conflicts.
      a.   true
      b.   false

Answers: 1.a, 2.b, 3.a

PART II: SELF-REFLECTIONS

The material presented in the text is often personally evocative and challenging. The questions below are designed to help students process and integrate their reactions to this information.

1.   Recalling the way you were raised by your parents, how would you characterize their parenting styles? What effect have their child-rearing practices had on your capacity for intimacy?

_____

_____

_____

_____

_____

_____

_____

_____

_____

_____

_____

_____

_____

_____

2.   Following Hartman (1978) or Minuchin (1974), referenced in
     the text, or any other format you may prefer, create a
     three or four generation genogram for your family-of-
     origin. Drawing from this perspective, describe both how
     your grandparents were parented and the child-rearing
     practices your parents experienced. In what ways have your
     child-rearing practices with your own children been similar
     or different?

PART III: SUPPORTIVE READINGS AND CASE STUDIES

The therapist's ability to embrace both positive and negative parts of clients' significant others is an important skill that helps clients become more self-accepting and self-affirming of what they experience as "flaws." The following case study highlights the importance of recognizing both sides of clients' ambivalent feelings toward significant others. This is an important integrative process that helps clients break pathogenic beliefs about having "flaws" that make them defective, unlovable, or shame-worthy. When therapists enable clients to experience and express both sides of their feelings, clients gain the ability to accept and integrate the full range of their emotions. Only by acknowledging both sides of their feelings, however, can clients ultimately resolve their conflicts.

Section A: Ambivalent Feelings: Case Study of Sheila

Sheila, a 15 year-old African-American female, had been sexually abused by her stepfather for almost eight years and had not, until now, directly told anyone about the molestation. There were, of course, many symptoms and problems related to the abuse, such as depression, overeating, expressions of anxiety about staying home alone with her stepfather, and several suicide gestures (which included taking over-the-counter pills and leaving the empty bottles on her parents' bathroom counter). However, it was not until Sheila established her first really close male friendship, and boys at school began to show interest in her and make "sexual" comments (such as commenting on how well she was "developing"), that she directly told someone about the abuse.

Sheila wished to be valued and loved and, out of that most human need, reached out to her family. This made her open and vulnerable to her stepfather, who made her feel full of shame and worthless. Sheila was faced with the poignant dilemma of needing support and affirmation from a stepfather who was exploitative and a mother who was invalidating. In response, Sheila adopted a compliant and caretaking stance in an attempt to be accepted by her family.

There are many aspects of this case we can examine, however, we are going to focus on the issue of intense *ambivalence.* Intensely ambivalent feelings occur for Sheila in this case, as well as for most clients who struggle with significant and enduring problems.

We believe that children's earliest relationships with significant others have tremendous impact on their developing self-concepts and serve as "templates" for future relationships. In Sheila's case, she internalized from her early relationships a sense of self suggestive of "worth" only when she was taking care of others and their needs. She was responded to positively only when she was compliant and pleasing others. This was, for example, evident in her relationship with her mother where she was often the caretaker, helping around the house, inquiring about her mother's needs, comforting her mother when she was sad, and rarely "burdening" her mother with her own needs. In her relationship with her stepfather, Sheila wanted to be cared for and special; she also wanted to avoid his anger at all costs. He would alternate between being genuine, kind, and helpful and repeatedly molesting her. When she resisted his advances, he became angry and withdrew from her. She felt that her only solution was to comply. Compliance was also congruent with her understanding of the religious teachings she had been exposed to that emphasized obedience to parents. In her desire to avoid her stepfather's anger and to maintain some level of relationship with him, Sheila submerged her own needs and voice. Over time, Sheila's ability to access and even express her own needs or feelings was profoundly muted.

Much of the initial phase of therapy was spent establishing a safe and affirming relationship in order to reduce Sheila's depression, including her suicidal ideation, sense of hopelessness, and self-condemnations. The process of engaging Sheila in therapy was intense, slow, and focused on empowering her to take the lead (e.g., therapist: "what would you like to talk about today"; "we can talk about whatever you choose to talk about"; "I appreciate it when you tell me that you are not ready to discuss that issue"; "you have not said much this session, is there something I have missed or which you wish to

discuss but are not sure how to approach"; and so forth). The

emphasis here was on closely tracking and together clarifying

her words, gestures, tones, and underlying messages and beliefs.

I was very careful not to repeat the relational process that had

occurred at home, especially the force/compliance pattern and

the invalidation. Thus, in order to create this reparative

interpersonal process, I broached potentially sensitive topics

tentatively and always respected her right to choose not to deal

with an issue at that time. I tried to convey, in words and

manner, that she was in control of the pace of therapy.

One way of undoing the invalidation Sheila had experienced

was to explore and together clarify the linkage between her life

experiences and her feelings. Her feelings of powerlessness,

hopelessness, and futility about her future made absolute sense

in light of her life experiences. Repeatedly, throughout the

course of treatment, I was trying to be validating of this

rather than minimizing the impact -- as many in her family had

done. This validation went a long way toward solidifying our

relationship and reducing her depressive symptoms. The

validation she experienced in our relationship provided the

security she needed to risk similarly asserting herself, asking

for help, and voicing her opinions in other relationships as

well. In other words, by enacting a different relational pattern in our interactions, Sheila felt understood, cared about, and became secure enough to be more authentic in other relationships.

It became clear early on that one of the ways Sheila tried to make sense of things -- given how profoundly she had been invalidated and the deep betrayal she had experienced -- was to conclude that she must somehow be "flawed." The sexual trauma had become incorporated into her sense of self, and she had begun to see herself as full of shame and unworthy of good things. It was thus important to question her pathogenic belief that she was "flawed", and to help her understand that bad things had indeed *happened* to her, but that *she was not "bad."* In doing this, I had to be "with" her (i.e., emotionally attuned and respectful) as she worked through her understanding of the moral and religious messages she had received. Toward accomplishing this treatment goal, my task was to listen and ask her to evaluate and clarify her understanding of these messages. In an effort to make overt her "cognitive schemas," I frequently asked questions such as, "Is that what you think (or believe or expect)?"; "I wonder what might make you think that way?"; "Can you think of other ways of thinking about this issue or come up with any other possible explanations?"; "Whose rule (or "should") is that?"; "Is that what you choose (or want) to believe?"; and so forth. Sheila clarified her pathogenic beliefs through this process of making overt her views of God, self, and others. The process was also intended to convey to her that I valued *her* thoughts, feelings, and beliefs and that I saw her as having worth. Again, this was an essential treatment goal given Sheila's history of invalidation and exploitation. My unwavering stance of respect and affirmation also began the process of providing her with a sense that this was a safe place where all her feelings and experiences could be discussed.

It was clear that Sheila approached most situations with great pessimism and had no positive future expectations. We began the process of magnifying her schemas by having her take a closer look at each of the assumptions she held (i.e., "it's my fault"; "there's something wrong with me"; "I can never do anything right"; or "nothing I do is ever good enough so there's no use in trying"). We explored the genesis of her schemas (i.e., her lack of affirming childhood experiences; the violation, betrayal, powerlessness, and stigmatization that resulted from the molestation; and the lack of protection from her mother) and their current functions (i.e., they were her way of making sense of events of her childhood and trying to instill order and some form of consistency where none had existed). We then evaluated their "reality" base (e.g., therapist: "where did these messages come from"; "do you really want to accept these as your own"; "are there other possible explanations"). This process was quite significant for Sheila. She began to ground her feelings in "reality" and began identifying areas where she could make choices and have some control. She was able to identify areas where she had been able to effect change (for example, she was no longer being molested by her stepfather), but she also recognized that the molest was not her fault because she was only a young child being coerced by a powerful significant other.

Over time Sheila's assumptions were less *automatic*. In the process (described above) of challenging her false beliefs, affirming her worth and dignity, and supporting her right to say no and be self-directing, Sheila took on the "questioning" model I had provided. She began to recognize her tendency to take on the negative characterizations made by others (e.g., "you are a very SELFISH child -- what would it hurt to drop the charges against your stepfather and just stay away from him" or "how is he going to support the other children if he is in jail"). Furthermore, she began to "talk through" who the statement belonged to and whose view it represented. Although statements such as these continued to have impact, they no longer were automatically accepted as "truth."

With regard to dropping the charges against her stepfather, Sheila agonized about the impact on her younger siblings if her stepfather was jailed. In the end, however, she stated that she did not want to recant the abuse charge since doing so would "be a lie." Although she had difficulty articulating this, she conveyed the sense that recanting would "confirm" forever her sense of being worthless. That is, it would continue the compliance with her stepfather, endorse the invalidation from her mother, and have her carry forward to future relationships this expectation and acceptance of denigration. In maintaining her stance and being supported in this by me and some of the people in her life, she experienced empowerment and acknowledged her self-worth. This was an especially delicate process given Sheila's family's value of maintaining the family-as-a-group at all costs. It was critical that Sheila, and not me as therapist, make the decision to keep the charge against her stepfather. My job here was to help her give a "voice" to all her feelings, including the family loyalty issues, the censure likely to result if she refused to drop the charges, and what dropping or following through with the charge would mean to her and her sense of worth.

Tracking the interpersonal process, Sheila's history of compliance with significant others made it imperative that I be highly cognizant of the extent to which I was directive. However, her level of depression during the initial phase also dictated that I be active. I tried to balance this by being active. This included providing information and asking questions such as "Is this what you want?," "Is this how you

feel?", and What do you think about?" I also tried to

"normalize" her experience and validate her feelings (i.e.,

"When you describe how hard you tried, it makes sense that you

would feel helpless.") Additionally, I used self-involving

statements such as "I'm feeling sad as you are telling me about

this." At the same time, I always maintained awareness of the

extent to which my activity might be construed as directive

(i.e., experienced as demanding or controlling in terms of her

internalized schemas). On many occasions I asked her directly

if she felt that I had a particular agenda or wanted her to

respond in a particular way, and we were then able to work on

this together. I also communicated my view that her needs,

feelings, and choices were important by acknowledging them

whenever she ventured to express them. In this way I was trying

to rework the role reversal and let her know that in this

relationship she would be responded to rather than have to take

care of me. I always kept in mind her "templates," which were

to please others, take care of them, and submerge her needs

below theirs.

During this phase, Sheila and I spent a great deal of time

talking about her stepfather's abuse of her. She recalled him

coming home during times when he knew she was alone or coming to

her room at night when the others were asleep. As we focused on

this, Sheila experienced an increase in intrusive thoughts and
was inundated with feelings of shame and rage. I explained to
her that this increase in pain was really a sign that she was
improving, and I used the medical analogies of chemotherapy and
surgery to explain how the treatment sometimes brings about a
temporary increase in distress and discomfort before healing
occurs.

Eventually, Sheila was able to express her rage at her
stepfather for violating her and making her feel so shameful.
During one particularly poignant session, Sheila began pounding
a pillow as she questioned why he had molested her. I quickly
set up a role-play, with several pillows representing her
stepfather, and encouraged her to express her feelings. As she
talked to him, her *ambivalent* feelings toward him emerged. In
addition to her rage and feelings of violation, she also
acknowledged many genuinely good things in their relationship
and how desperately she had wanted to be loved by him. At the
end of the role-play, Sheila was able to talk about some of his
characteristics that she had valued. For example, he would help
her make breakfast or praise her for some accomplishment --
statements that had made her think of him as a caring father.
She struggled to understand how he could be so kind at times and
then be so profane at times. Sheila's ability to acknowledge
both the positive and negative aspects of her stepfather was

crucial in helping her develop empathy for herself. An important part of this was realizing that life contained many gray areas and that people had good and bad parts to them. Additionally, she recognized that people, including herself, could be loved despite their flaws. Grasping this helped Sheila to not remain stuck in her rage toward her stepfather. Further, a significant paradigm shift occurred as Sheila began to feel the compassion for herself that I had been feeling for her.

Sheila continued with individual and group therapy and terminated when she prepared to go off to college. Treatment was successful. Sheila joined a new, supportive church, her grades improved, and she created a more affirming peer group and expanded her social support network.

Throughout treatment, I struggled with my own countertransference issues. I was furious at her stepfather for molesting her, of course, but was also outraged at her mother's continued failure to protect her. I had to work hard and sought consultation in order to prevent my negative reactions toward her parents from impeding my ability to hear and affirm her genuinely positive feelings toward both of them.

Sheila's case highlights the importance of recognizing both sides of clients' ambivalent feelings. This integrating process is a bridge to help clients break the common pathogenic belief that having one "flaw" makes them defective and shame worthy.

Thus, when therapists express and focus only on their negative feelings toward caregivers who have hurt their offspring, they curtail the clients' ability to come to terms with their full range of emotions toward those individuals, which routinely include loving feelings as well. When the therapist fails to embrace both the positive and negative parts of clients' relationships with significant others, clients are prevented from accepting and affirming themselves and their own "flaws" or weaknesses that falsely led them to conclude that they were unworthy. This also then results in clients being unable to affirm their positive characteristics. Since everyone has positive and negative aspects, the ability to acknowledge and embrace all characteristics of one's self is a core feature of good psychological health.

PART IV: JOURNAL ENTRY

Use the following section to further integrate the material presented, to address unresolved questions or issues, and to further explore your personal reactions.

_____

_____

_____

_____

_____

_____

_____

_____

_____

_____

_____

_____

_____

_____

CHAPTER SEVEN: INFLEXIBLE INTERPERSONAL COPING STYLES

PART I: STUDY GUIDE

Part I will help students prepare for course examinations. It consists of three sections: a list of key terms to define; essay questions for central concepts; and sample multiple choice and true/false questions.

Section A: Key Terms

Provide a one or two sentence answer for the following key terms.

DEVELOPMENTAL NEEDS _____

_____

_____

COMPROMISE SOLUTIONS _____

_____

_____

ENVIRONMENTAL BLOCK _____

_____

_____

COGNITIVE SCHEMAS _____

_____

_____

INTERPERSONAL COPING STYLE _____

_____

_____

RISING ABOVE _____

_____

_____

ANXIETY _____

_____

_____

INTERNALIZED TAPES _____

_____

_____

MOVING TOWARD _____

_____

_____

MOVING AWAY _____

_____

_____

MOVING AGAINST _____

_____

_____

NEUROTIC PRIDE SYSTEM _____

_____

_____

GENERIC CONFLICT _____

_____

_____

HONORING INTERPERSONAL COPING STYLES _____

_____

_____

ACKNOWLEDGING CONFLICTS _____

_____

_____

SENSE OF ENTITLEMENT _____

_____

_____

TYRANNY OF THE SHOULDS _____

_____

_____

NARCISSISTIC WOUND _____

_____

_____

EMOTIONAL RELEARNING _____

_____

_____

Section B: Key Concepts

    Consider each of the following key concepts. Be prepared
to write a two to three paragraph essay for each of the
following questions.

1.   Describe one way that parents could block a child's
     developmental needs and how this may shape the child's
     sense of self. How might this relate to the interpersonal
     coping strategies that the child develops?

_____

_____

_____

_____

_____

_____

_____

2.   Consider the different characteristics of each of
     Horney's three interpersonal coping styles. Describe how
     each might be expressed in the therapeutic relationship.

_____

_____

_____

_____

_____

_____

_____

3. Which of Horney's three interpersonal coping styles may be most difficult for you to respond to? Suggest effective and ineffective therapeutic interventions for this interpersonal coping style.

_____

_____

_____

_____

_____

_____

_____

_____

4. Explain why the "rising above" side of compromise solutions inevitably fails.

_____

_____

_____

_____

_____

_____

_____

_____

Section C: Sample Multiple Choice and True/False Questions

Answer the following sample multiple choice and true/false questions.

1.   Children are most likely to associate _____ with their own unmet developmental needs.
     a.   anxiety
     b.   anger
     c.   fear
     d.   compulsion

2.   According to Horney, clients who rise above their unmet needs through aloofness and glorify this stance by feeling self-righteously self-sufficient, have a characterological interpersonal style of:
     a.   moving toward.
     b.   moving inward.
     c.   moving against.
     d.   moving away.

3.   Compromise solutions clients adopt are doomed to fail because the client cannot rise above the original conflict.
     a.   true
     b.   false

Answers: 1.a, 2.d, 3.a

PART II: SELF-REFLECTIONS

The material presented in the text is often personally evocative and challenging. The questions below are designed to help students process and integrate their reactions to this information.

1.   Which interpersonal coping style do you primarily employ (moving away, moving toward, moving against)? How does this work for you? Alternately, what does this coping strategy cost you internally? Thinking back on your family of origin, why do you think you developed this strategy, and how was it once adaptive?

_____

_____

_____

_____

_____

_____

_____

_____

_____

_____

_____

_____

_____

2. Think of two or three significant people in your life.
    What do they do, if anything, to maintain their sense of
    "specialness" or neurotic pride? What do those mechanisms
    tend to evoke in you in relation to these people?

_____

_____

_____

_____

_____

_____

_____

_____

_____

_____

_____

_____

_____

_____

_____

_____

_____

3. Which interpersonal coping style is most difficult for you
    to respond to? What does this particular style tend to
    evoke in you, and how do you tend to respond initially?

    _____

    _____

    _____

    _____

    _____

    _____

    _____

    _____

    _____

    _____

    _____

    _____

    _____

    _____

    _____

    _____

    _____

PART III: SUPPORTIVE READINGS AND CASE STUDIES

The following case study of Hap Loman was taken from "Death
of a Salesman," starring Dustin Hoffman. It illustrates how
students can use and apply the Case Conceptualization/Treatment
Planning format presented in Appendix B of the text to an actual
case. Readers are encouraged to view the Dustin Hoffman version
of this play prior to reading the case study.

Section A: Case Conceptualization Guidelines: Case Study of
                  Hap Loman

1. Formulate the Problem(s)

Hap Loman is a 36-year-old Caucasian male who has been married for
six years and has one son. Hap and his family live with his widowed
mother in the family home in a small urban lower-class neighborhood
in Pennsylvania. Hap has recently sustained some serious losses in
his life. He was dismissed from his position as a salesman in a
medium-sized manufacturing company for having an affair with his
boss's wife. The revelation of the affair caused his own wife to
threaten to leave him and take their child with her. The possibility
that his life will be disrupted has prompted Hap to seek therapy.
Hap has several issues to explore, including his tendency toward
excessive drinking, a DUI, and a domestic violence incident. Hap
appears to be depressed by these events and has mentioned to his wife
and his mother that suicide would be a solution to his problems.
Hap's father committed suicide 6 1/2 years ago, an event his mother

still hasn't recovered from. When Hap mentioned suicide, his mother threatened to kick him out of the family home rather than "go through that again." Hap relates the details of his father's suicide without affect and speaks of his wife and mother in disparaging, condescending terms. He blames his wife for his affairs, his mother for his failure to continue in school, and his father for committing suicide. He believes his life would have been better if only these people hadn't stood in his way.

Hap has a similar view of co-workers and friends. He appears puzzled that his co-workers and boss don't "get" his superior intelligence and specialness. He frequently disparages each of them in detail, gloating over their inadequacies, while extolling his own virtues and bemoaning the unfairness of people.

Historically, Hap has not seen the need for psychotherapy and would not be seeking help now if it weren't for the pressure to do so from his mother and his wife. The significant familial factors that make seeking help difficult for Hap are the grandiose modeling by Hap's father, Willie, and his parents' neglect of him in favor of his older brother, Biff.

When I began seeing Hap, I initially held a working diagnosis of an Adjustment Disorder or an Affective Disorder. However, after seeing Hap for three weeks, I have noticed that he has many of the characteristics of someone with Narcissistic

Personality Disorder: (1) an exaggerated sense of self-importance; (2) a need to be in the limelight; (3) fragile self-esteem; (4) preoccupation with how others perceive him; (5) a deep disregard for the feelings of others; (4) a sense of entitlement; and (7) lacks empathy. Since Hap's symptoms are pervasive and longstanding, I believe Hap's diagnosis should be: Axis I - Dysthymic Disorder (300.4), and Axis II - Narcissistic Personality Disorder (301.81).

2. Treatment Focus

Hap's core conflicted affect and corresponding self schema is one of feeling shamefully inadequate and expecting to be subtly ignored or overly rejected. Hap wants to be loved, valued, and most of all, noticed. He was ignored in favor of his brother Biff by both of his parents and has since felt that he doesn't quite measure up. He is unsure what yardstick is being used to measure his worth, so he pretends that he is good at everything. He believes that if he were as successful as he deserves to be people would love him the way Biff was "loved" and his life would be fine. Instead of feeling loved, he feels that people stand in his way, take credit for his accomplishments, and withhold the admiration he believes he deserves. Hap feels alternately aggrandized and degraded and expects to be ignored, devalued, and unloved, so he blames other people for not seeing his true worth. This translates into contempt for others ("they are too stupid") and self-contempt (shame regarding his feelings of inadequacy and unworthiness or of believing he is unlovable). Hap uses blaming, flattery, empty promises, and grandiosity as interpersonal coping strategies to "rise above" and manage his pervasive relational problems. These moving-against coping strategies tend to elicit feelings of being manipulated and, as a result, others tend to feel used and resentful toward Hap.

Hap suffers and defends against a painful narcissistic wound to his most basic sense of self. When he is at risk of having this core defect or flawed sense of self revealed, Hap holds away the shame by placing unrealistic demands on himself and others. When these aggrandizing attempts to restore his sense of adequacy fail (as they repeatedly do), Hap feels self-loathing and contempt for others -- often striking out verbally and criticizing others harshly. Occasionally the shame-rage cycle escalates to physical aggression. Hap wants to be loved, but experiences his relationships as superficial and himself as empty. Never feeling sufficiently loved or appreciated, he compulsively has affairs -- looking for the adoration he craves and can temporarily elicit. Experiencing himself as being inadequate, not enough, and readily forgettable, Hap expects others to withhold their time, approval, affection, and

attention from him and give them to someone else. Consequently,
he feels shame about himself and contemptuous of others.

Hap's interpersonal strategies involve telling people how
great he is (an attempt to "rise above" his painful unmet
needs). He believes that if people could just understand how
special he is they would give him the love he deserves. Thus
Hap embellishes his part in the stories he tells to make himself
look like the hero and others look either stupid or inadequate.
This represents Hap's strategy of *avoiding* his feelings of
worthlessness. Hap also compliments others in the hope of
gaining their admiration, support, and collaboration. Although
Hap's charm is often successful in the shortrun, this approach
leads others eventually to feel manipulated, taken advantage of,
and degraded.

One focus for treatment is to make overt Hap's maladaptive
interpersonal coping strategies and what these are defending
against. This is touchy because of his fragile self-esteem, his
grandiosity, and his need to be seen as "special" and perfect.
If I can convey a consistent support for Hap and the personal
qualities that he genuinely possesses without buying into the
exaggerated fantasies that he pretends to have, he may begin to
develop a more realistic or balanced sense of self between the
two extremes of his defensive grandiosity and his self-contempt.
Initially this will evoke shame in him, and I will need to

provide a safe "holding environment" and help him tolerate these profoundly threatening feelings. If I can achieve this challenging treatment goal, sadness or grief over what he missed growing up will emerge as he feels the safety to begin surrendering his compensatory illusions of "specialness" and accept a self that is more human than extraordinary.

3. Developmental Context.

Hap grew up in a lower-middle class neighborhood with an older brother, Biff, who was the family hero. His mother was a homemaker, and his father, Willie Loman, was a traveling salesman who was on the road for weeks at a time. Willie had the "gift of gab" and liked to entertain his family with stories of the road that he freely embellished to enhance his self-esteem and sense of importance. In reality, the family was nearly impoverished, borrowing regularly from a neighbor in order to make weekly payments on household appliances and monthly mortgage payments. Willie believed making a "good appearance" was the most important factor in creating a successful life. Therefore, he felt compelled to enhance his own appearance by fabricating accomplishments and embellishing ordinary achievements to the extent that soon they bore little resemblance to the truth. By attending to the superficial components of life and ignoring the more substantial building blocks of personality development such as honesty, hard work, humility, and respect for others, Willie was instrumental in the formation of his son's narcissistic character. Willie taught his boys they were "special" and that they didn't have to follow the same rules that applied to other boys. Willie placed no emphasis on working hard for good grades, honesty, or integrity. He talked about and reinforced, instead, the "big play" of the football game, the "big sale" on his road trip, and the "big score" in life of financial riches. Mrs. Loman colluded fully in this deceptive family mythology, and the boys internalized this value system as their own. Each came to believe he was special and not bound by the rules and limits of society. Biff demonstrated this by stealing repeatedly, and Hap by arrogantly sleeping with the wives of other men who held authority over him. They took what they wanted with little regard for the feelings of others or the consequences of their actions, including the self-destructive impact they were having on their own lives.

A second family dynamic was the pivotal relationship between Biff and Willie. Willie formed a collaborative cross-generational alliance with his son, Biff, while verbally abusing and devaluing his wife. This made Biff feel both special and anxious. He felt special at first when he viewed his father through the eyes of a naive boy, and anxious (and furious) later when he found his father with a woman in a hotel room. Hap and his mother were the "audience" for this primary relationship, lending credence to it by their acceptance of it. Hap was pushed aside and ignored by Willie and Linda -- developing early on a sense of "not being good enough" and not being seen. When Biff left the family to move out West, Hap tried to step into the spotlight, but it had a dimmer bulb. Looking for an identity when his own authentic interests and abilities were ignored, he took on his father's profession and his father's braggadocio style. He was never able to realize the "special" place that Biff had held, however, and his jealousy and resentment ate away at him. He acted out of his "dark side" by vindictively "getting back" at people whom he felt shamed or belittled by. He began sleeping more frequently with married women, specifically the wives of his boss and co-workers. He also devalued his supervisors and co-workers verbally, disparaging their accomplishments while aggrandizing his own.

Hap's social supports consisted of his "nagging" wife and his idealized but guilt-inducing mother. Both were pushing him to make changes in his life. However, because of his inability to reveal himself and his lack of feeling valued or affirmed in any relationship, these familial relationships were superficial, lacked emotional intimacy, and did not hold the salience necessary to motivate real change.

The most central pathogenic belief Hap holds about himself
is the two-sided coin of his "special status" in the world,
coupled with his underlying experience of himself as
insignificant and worthless. He presents himself as having
special talents and abilities, and feels that others don't
appreciate him the way they should. Hap exaggerates his
accomplishments and then believes the lies, just like his
father. In sum, Hap's basic sense of self is that of being
defective and unworthy, which he tries to rise above via a
grandiose, moving-against coping strategy.

4.   Therapeutic Process

The first issue of therapy was to deal with Hap's suicidal
ideation. He assured me that he was not serious about taking
his life, that he was just "blowing off steam." However, we
negotiated a "no-suicide" pact for the length of his therapy. I
also did an informal suicide assessment approximately every
other session to measure where Hap was on this important issue.
I feared that further failures to be special and grand could
evoke intense shame, alcohol abuse, and significant potential
for suicidal escape. Second, I wanted to build a collaborative
working relationship with Hap. This was not easy to do, of
course, because Hap liked to keep our interaction on a
superficial level. He complimented me frequently, always
noticing a new dress or hairstyle. I wondered if Hap was trying

to convey that he needed or wanted me to compliment him and make him feel special. I made process comments that were designed to make overt his behavior that seemed to be aimed at "seducing" me, such as: "I think something important may be going on right now. Can we talk together about what just happened between us? As you are complimenting me, I am wondering what you are wanting me to feel or do or say? How do other women in your life usually respond when you act this way toward them? Where does that leave you?" Hap denied that he had any "intentions" and acted offended that I could not take a "simple compliment" for what it was. He questioned my competence and brought up the fact that I was a graduate student counselor and not very experienced. I explained myself further by making the point: "I know that people don't usually talk together this forthrightly. But I think that if we could talk about our relationship and try to understand what goes on between us it could help us understand what's been going wrong with your wife, your boss, and others." I appreciated that Hap's seductive style was part of his "rising above" strategy to ward off his shame-based sense of self by feeling grandiose, powerful, and special but also recognized how dearly this coping style could cost him.

Hap also brought his conflicts with others into the therapeutic relationship by attempting to elicit my support of his "special" status. This was done by trying to change the fee structure, coming late to appointments, and by canceling

appointments at the last minute. Furthermore, he tried to see

me at unscheduled times by showing up at my office and expecting

me to "fit him in." Recalling that Hap did not have consistent

or realistic limits as a child, I kindly but firmly let Hap know

that our appointment times and fees were set and needed to be

negotiated in order to be changed. In this way, I also

attempted to let Hap know that I understood he had a need to

"rise above" his generic conflict of feeling unworthy and

devalued in an attempt to be perceived by me as special. I knew

that he experienced shame as a result of my not buying into the

need, but I let Hap know by my consistent responsiveness and

nonjudgmental concern that I saw and experienced him "as he was"

and that I liked and accepted many parts of him. I hoped that

this would gradually help free him to acknowledge and accept his

own authentic self.

I also tried to convey that I understood why he had needed

this interpersonal coping strategy given his family history. I

wouldn't comply or "charge back" when he became anxious and

employed his angry, derogatory, competitive, or demanding

interpersonal defenses. I told him he didn't need to be

"special" with me in order to be liked, that he could be himself

with me and still be accepted and cared about. This took some

time in therapy, but over several months Hap increasingly came to trust that I would not disparage or disapprove of him when he was vulnerable and otherwise imperfect.

Although Hap was not very aware of it, he desperately wanted everyone, including me, to admire him inordinately. This wish conflicted with his true expectations, that he was basically unlovable. As a result, when I conveyed to him that I accepted him as he was, this was initially very foreign to him and made him anxious. My acceptance evoked the shame and pain of his lifelong emotional deprivation, which he defended against by sexualizing our relationship, idealizing me in a bid for mutual aggrandizement, or denigrating my ability to help him. These process discussions were ultimately very productive for him as he began to link these maladaptive relational patterns he was enacting with me to other relationships in his life. As we explored when and why he employed these coping strategies with me, and with others, it reduced his need to enact these well-worn relational patterns that kept leaving him anxious, depressed, and lonely.

Extending these connections between our interaction and his presenting problems, Hap tended to elicit in me, as he did in others -- feelings of being manipulated. Hap's need of constant reassurance about being special and his insincere compliments make it difficult for me to feel connected to him in a genuine way. When I feel he is trying to manipulate me, I tell him

gently how I am perceiving his compliments and how it feels
uncomfortable or even controlling to me. I let him know that
being flattered does not influence how I feel about him as a
person and that he has many qualities that I do value. I then
tell him what these are -- such as his intensity, sense of
humor, and genuine enthusiasm for life. I also explain, as
tactfully as I can, that I would rather he not flatter me
anymore, especially because it feels like replaying a
problematic old pattern for him.   I will still like him, and in
fact I will respect him more, if he is honest with me. I also
acknowledge that these are some of the things he missed in his
childhood -- the value of authenticity in relationships and of
being honest and direct with people. In these ways some of
Hap's pathogenic beliefs and faulty expectations are
disconfirmed, particularly the belief that I could only value
him if he was obsequious and/or extraordinary.

In sum, Hap tends to distort and misperceive me by
continually attempting to change the nature of our relationship.
He tries to be my friend, my lover, and my star. In turn, he
wants me to be the audience for his grandiose fantasies, the
recipient of his office gossip, and of his flattery. In each
instance I stick to our professional relationship, bringing Hap
into the present with respectful but reality-based feedback. I
exclude gossip by not attending to it, flattery by telling Hap

that I don't need him to flatter me, and using self-involving
statements when I feel manipulated by him.

5. Goals and Interventions

One major goal of Hap's therapy is to allow him to become
comfortable with his authentic self or "own voice." I am
attempting to accomplish this goal by giving Hap a corrective
emotional experience regarding his inner fantasy life of
performing "on stage" in every interaction to admiring others,
fantasizing, his embellishment of his accomplishments, and his
interpersonal style of alternately complimenting or disparaging
me. When Hap applauds himself or me, I let him know that it
isn't necessary to make either one of us bigger than life or
more than we really are. I am seeing Hap as he is and value him
(and myself), human defects and all. Hap has much shame
connected to who he is and his perceived inadequacies. I
repeatedly 'see' Hap, letting him *know* that I see him (in his
ordinariness or just plain humanness) and that he still matters
to me. I also try to accomplish these goals by helping Hap see
that denigrating his co-workers and lovers while glorifying
himself is at best maladaptive and hurts him far more than he
has realized. We explore where and why he originally learned to
cope in these ways, and together we explore better ways to
define himself and respond to others. We make overt the fact
that, as a child, he did indeed need to be special in order to

"matter" in his family, but that this interpersonal coping strategy is now no longer needed.

Long-term goals for Hap include internalizing my consistent and more balanced regard for him and generalizing that respect and regard to others. I believe that as I help contain the core conflicted affect of shame and, in turn, his sexual and alcohol-related defenses against experiencing it, Hap will feel more internally secure. As he repeatedly finds that I value him, he will become increasingly capable of realistically valuing himself and will no longer need to devalue others to bolster himself. That is, because I do feel empathy for Hap and his experience, I am providing or modeling the attitude I would like him to adopt toward himself and others. It will not be easy for him to internalize this, however, because getting some of what he has always wanted will evoke the full impact of his childhood deprivation, and the shame, pain, and expectations of further rejection that will be evoked will require repetitious working through.

6. Impediments to Change

Hap likely will not stay in therapy for long unless I can convey to him that I value and feel for his authentic self -- and not the inflated or grandiose persona he tries to portray. Initially, I need to align myself with Hap and let him know that I am not a threat to him. I achieve that by behaviorally demonstrating that I value him, grasp the very big problems he has been struggling with all of his life, and that I will not be dominating or dominated. Thus I need to anticipate his contempt and criticism of me, which are likely to arise at any time, and be prepared to respond nondefensively. This will be especially important when he begins to experience the shame that arose from his familial experiences of being ignored, denigrated, and controlled, all of which combined to make him feel unworthy. In Hap's experience, it was his brother Biff who received all of his father's attention and love. Hap was ignored and pushed aside. This relational template is going to be activated many times in therapy, but I need to be able to name it or find other ways to work with it so that we do not recapitulate his generic conflict (i.e., by being late for an appointment with him, giving my full attention to the client before him when he arrives, rearranging his appointment to accommodate another client). Hap must feel secure enough with me to be able to begin to let down some of his defenses (such as his boasting or sexualizing), which will entail my being emotionally present when we are together and not buying into (or reacting punitively) to either his aggrandizement or degradation of me. This corrective emotional experience will gradually engender Hap's respect, and we will begin living out a different kind of relationship that he has not experienced before. For example,

when Hap speaks in disparaging terms about his co-workers, I
will try to generate more realistic hypotheses about them
without eliciting Hap's fear that I think they are better than
he is (as both of his parents did). If Hap thinks I prefer
another over him, it will elicit anxiety about not being good
enough or that I see his core defects and scorn him. I am aware
that this will probably cause him to criticize me or to
embellish himself and his accomplishments. I need to be alert
for these signals, understand them as defensive maneuvers to
minimize anxiety, and help Hap appreciate that these coping
strategies made sense in the context of his history but are no
longer necessary with me or adaptive with others. In other
words, these modes of defense make sense for Hap because Willie
taught him that what matters most is how one is _perceived_ by
others rather than how one actually "_is._" Although this life-
long coping style has caused Hap problems in every arena of his
life, for him to change this will violate family rules and cause
him to feel profoundly guilty and disloyal to his parents. If I
fail to appreciate Hap's loyalty to his parents and attempt to
show them up, he will terminate prematurely.

Hap was ignored and pushed out of the spotlight in his
family of origin, while his older brother was given all the
adulation that Hap mistook for love. This old relational
scenario could be reenacted, and he could leave therapy

prematurely, if I don't consistently demonstrate in word and deed that I see and hear him as he truly is and do not mistake the real human being for either the grandiose mask or the shame-worthy expectation. Thus, I acknowledge his courage when he is honest and straightforward with me or others and convey that he is good enough as he is and doesn't need to embellish more for me to value him. My own countertransference may come into the therapy if: (1) I am not watchful about Hap's wanting "special status" regarding scheduling changes and fee reductions; (2) I become defensive over Hap's criticism of me, my student status or my inexperience; and (3) I buy into his flattery or idealization of me. As these countertransference possibilities develop, I must seek supervision as reenactments with Hap would escalate quickly and he could avoid conflicts readily by terminating impulsively.

PART IV: JOURNAL ENTRY

Use the following section to further integrate the material presented, to address unresolved questions or issues, and to further explore your personal reactions.

_____

_____

_____

_____

_____

_____

_____

_____

_____

_____

_____

_____

CHAPTER EIGHT: CURRENT INTERPERSONAL FACTORS

PART I: STUDY GUIDE

Part I will help students prepare for course examinations. It consists of three sections: a list of key terms to define; essay questions for central concepts; and sample multiple choice and true/false questions.

Section A: Key Terms

Provide a one or two sentence answer for the following key terms.

ELICITING MANEUVERS_____

_____

_____

TESTING BEHAVIOR_____

_____

_____

TRANSFERENCE REACTIONS_____

_____

_____

RELATIONAL REENACTMENTS_____

_____

_____

FORMULATING WORKING HYPOTHESES _____

_____

_____

PASSING TRANSFERENCE TESTS_____

_____

_____

CLIENT RESPONSE SPECIFICITY_____

_____

_____

ASSESSING CLIENT REACTIONS _____

_____

_____

REENACTMENT OF CLIENTS' CONFLICTS _____

_____

_____

ENCODED REFERENCE TO THERAPIST _____

_____

_____

TRACKING THE THERAPEUTIC PROCESS _____

_____

_____

CONCEPTUALIZING TRANSFERENCE DISTORTIONS _____

_____

_____

ENMESHMENT VS DISENGAGEMENT _____

_____

_____

TWO-SIDED NATURE OF CONFLICT _____

_____

_____

AFFIRMING AMBIVALENT FEELINGS _____

_____

_____

Section B: Key Concepts

Consider each of the following key concepts. Be prepared to write a two or three paragraph essay for each of the following questions.

1.   Discuss the concept of the two-sided nature of clients' conflicts and how this is often immobilizing for clients. Suggest effective and ineffective interventions for helping clients explore both sides of their conflicts and resolve their ambivalent feelings.

_____

_____

_____

_____

_____

_____

_____

2.   Discuss Sullivan's concept of clients' "eliciting" behavior. How might these maneuvers serve to protect clients from suffering further pain and anxiety? What are effective and ineffective ways of dealing with this behavior within the therapeutic relationship?

_____

_____

_____

_____

_____

_____

_____

3.   Recall Weiss's concept of "testing" the therapist. How
     could a therapist use this behavior to help clients
     resolve conflicts?

_____

_____

_____

_____

_____

_____

_____

_____

_____

4.   "Client response specificity" is a significant factor in
     responding to clients' testing behavior or in providing a
     corrective emotional experience. Why is it so essential?

_____

_____

_____

_____

_____

_____

_____

5. Discuss how counselors can use clients' transference
    reactions to better conceptualize their clients' conflicts
    and to guide subsequent treatment plans.

    _____

    _____

    _____

    _____

    _____

    _____

    _____

    _____

    _____

6. Discuss the typical ways in which beginning counselors
    become enmeshed or disengaged with their clients. What
    can these counselors do to obtain a more effective or
    optimum interpersonal balance?

    _____

    _____

    _____

    _____

    _____

    _____

    _____

Section C: Sample Multiple Choice and True/False Questions

Answer the following sample multiple choice and true/false questions.

1.   When clients are able to successfully utilize eliciting maneuvers with their therapists, they:
    a.   attain protection from the generic conflict and <u>no</u> effective change occurs.
    b.   attain protection from the generic conflict and change effectively within the therapeutic relationship.
    c.   are not sufficiently motivated to change in therapy.
    d.   must explain what this means.

2.   Therapists "pass the client's tests" when they:
    a.   respond to all of their clients in a consistent and reliable manner.
    b.   respond in a way that disconfirms the old relational patterns or templates.
    c.   comply with the client's requests and expectations.
    d.   set limits on the client's eliciting maneuvers.

3.   The healthier a person is, the more intense and pervasive the transference reactions.
    a.   true
    b.   false

Answers: 1.a, 2.b, 3.b

PART II: SELF-REFLECTIONS

The material presented in the text is often personally evocative and challenging. The questions below are designed to help students process and integrate their reactions to this information.

1.   It is often very intimidating for beginning therapists to track all of the complex and multifaceted therapeutic processes simultaneously. Clarify your own learning pattern by recalling other challenging new tasks you have attempted. In those situations, how did you initially feel about yourself? How did you tend to respond to new demands, and what helped you eventually master the new tasks? How long did those processes take you, and how did you feel upon completion?

_____

_____

_____

_____

_____

_____

_____

_____

_____

_____

_____

_____

2.   Thinking of your own relational templates and response
     patterns, identify the type of eliciting maneuvers or
     testing behaviors that will be most difficult or
     challenging for you to respond to. What will you initially
     tend to say and do when a client uses these with you?

     _____

     _____

     _____

     _____

     _____

     _____

     _____

     _____

     _____

     _____

     _____

     _____

     _____

     _____

     _____

     _____

     _____

3. Recall a specific situation in which you felt pushed or
    pulled to act in an inauthentic manner. What did this
    evoke in you? Did you resolve this dilemma? Can you
    identify responses from others that helped you find your
    own voice and helped you to behave more authentically?

    _____

    _____

    _____

    _____

    _____

    _____

    _____

    _____

    _____

    _____

    _____

    _____

    _____

    _____

    _____

    _____

    _____

4. Recall a situation in which you felt intensely ambivalent
    about an important decision. Identify responses that
    helped you resolve this conflict and contrast them with
    responses from others that were not facilitative.

_____

_____

_____

_____

_____

_____

_____

_____

_____

_____

_____

_____

_____

_____

_____

_____

_____

_____

PART III: SUPPORTIVE READINGS AND CASE STUDIES

The following case study of Linda Loman was taken from the film adaptation of Arthur Miller's great play, "Death of a Salesman," starring Dustin Hoffman and John Malkovich. As in Chapters seven and nine, this case study illustrates how students can apply the Case Conceptualization/Treatment Planning format presented in Appendix B of the text to an actual case. Readers are encouraged to view the Dustin Hoffman version of this play prior to reading the case study and to consider how they might conceptualize Linda's problems.

Section A: Case Conceptualization Guidelines: Case Study of Linda Loman

1. Formulate the Problem(s).

Linda Loman, a 60 year-old, white female came in to see me after being referred to therapy by her physician. She complained of feeling tired, yet she was unable to sit and relax. Linda also stated she was having trouble falling asleep at night due to her "constant worry" over her two adult sons, Biff and Hap. Two years ago Linda's first husband committed suicide, and she remarried one year later. Linda stated that she believed her current husband was a hard worker who was "slightly impatient" due to the stress of his work. She described him as a "wonderful man" who was misunderstood by his supervisors and co-workers, as well as by her sons. According to Linda, her sons Biff and Hap were "good boys" who were also struggling with their careers. Linda stated she has always worried a lot, especially about her family. Due to Biff and Hap's recent business failure, they had moved back home with their mother and stepfather. Neither son was dealing well with the failure, and their frustration had added to the tension in the home. Just like her first husband, Linda's second husband lacked interpersonal skills in his work setting. He was also openly resentful and hostile toward Biff and Hap because of the financial burden their presence added to the home. Linda indicated that she had been desperately trying (to no avail) to "keep the peace" and make her family "happy." She was feeling as though her "world was falling apart," and she didn't know what to do to "fix" it.

Linda's approach to her problems was passive and included denial of her problems and feelings. She also believes that it's wrong or "disloyal" to discuss these with others. Linda tries to keep peace in her family by pulling her sons aside and begging them to "get along" with their stepfather.

Linda's belief that it's wrong to talk about her problems, and her tendency to deny them, has prevented her from seeking help until now. Her first husband, Willy, also believed in maintaining secrecy and refused help, which contributed to his suicide. Linda has not told her family about coming to see me.

Since my fee is being paid by a state insurance fund and does
not come out of the household budget, she is able to keep her
therapy a secret.

Linda's blue-collar family of origin was traditionally
patriarchal and viewed the husband as the head of the household.
Her mother was "submissive" to her father and modeled this
female role for Linda. Currently, Linda and her family have few
financial resources. Linda wants her family to "look good" and
appear more prosperous than they are. This is especially
evident in her awkward interactions with others who are more
financially secure. Although she denies herself, she makes sure
that her husband and sons are well dressed. Linda's guilt over
having things for herself, shame over having her unmet emotional
needs revealed, and underlying feelings of inferiority
contribute to her tendency to "martyr" herself and induce both
guilt and shame in others. Using this interpersonal coping
strategy of martyr, in which she "denies" herself and presents
as the peacemaker who puts others first, Linda effectively
manages to manipulate others to meet her needs without having to
"own" her requests.

Linda's symptoms suggest an Axis I diagnosis of Generalized
Anxiety Disorder (300.02). This includes: (1) her inability to
control her excessive worry over her sons, husband, and
finances; (2) her symptoms of fatigue, restlessness, and sleep

disturbance; and (3) her low social functioning. In addition, Linda reports feeling depressed "most of the time," which has been continuing for as long as she can remember. In conjunction with her fatigue, low self-esteem, and insomnia, these symptoms suggest a co-morbid Axis I diagnosis of Dysthymic Disorder, Early Onset (300.4). Linda also presents with passive-aggressive features. She is rarely direct about her wishes or needs, portrays herself as "self-sacrificing," and fulfills the role of a "martyr," yet underlying this is an ascerbic and contemptuous tone that indirectly demands compliance.

2.   Treatment Focus

Linda wants to be loved, accepted, and cared for by her husband and sons -- and by me. However, she is afraid that if she allows herself to have needs or be responded to that those she depends on will, in sequence, denigrate her emotional needs, reject her, and leave. As a result of this relational template, she denies having any needs of her own and copes by putting others' needs ahead of her own in a "martyrish" way. Her defense against overtly acknowledging her own genuine needs or authentic voice is also apparent in our relationship when she tries to "take care" of me by being a "good" client and telling me what she thinks I want to hear. I experience these efforts as inauthentic, and they keep me from getting to know the "real" Linda.

As noted above, Linda's templates lead her to expect to be criticized, discounted, and abandoned by others if she expresses what she really wants or believes. That is, if she overtly acknowledges her needs and tries to get these met, they will not be met and she will be both painfully disappointed and shamefully rejected. Consequently, she is indirect and manipulative with me regarding her needs, preferences, and wishes. She also tends to paint a rosy picture by talking as if everything will be "fine." For example, Linda will admit that she is worried about her sons' business failure, but she immediately glosses over her worry by saying she is sure they will find a new venture soon. It is probably too threatening for Linda to face this disappointing situation realistically because it taps into her feelings of inferiority and the discrepancy between her family's *actual* and *portrayed* level of functioning and prosperity.

Linda's inability to accomplish the superhuman task of "fixing" her family's problems leaves her feeling inadequate, helpless, and hopeless. She expends a great deal of energy maintaining the pathogenic belief that if she just keeps trying everything will work out. When things don't work out, however, her deep feelings of helplessness emerge, and she covertly communicates that she expects me to "fix" her problems and is subtly disparaging of me for seemingly failing her.

Linda experiences several conflicting affective states in the course of her everyday life. For example, on Monday mornings she anxiously hopes that her husband will have a "good week" in sales to make enough money to pay the bills. As the week progresses, however, Linda feels hopeless and becomes more depressed as this does not materialize and she covertly expresses disappointment and anger through subtle criticisms, demands, and belittling innuendoes. Continuing this sham of conflicting emotions, Linda feels secure and proud that her husband "still has a job." When he boasts of his good reputation as a salesman, however, she feels ashamed because his boisterous bragging embarrasses her in front of their neighbors. Since Linda draws her identity from her role as wife and mother, her husband's socially inappropriate behavior makes her feel embarrassed and alienated from friends and neighbors.

Because Linda portrays herself as a "martyr" who always puts others first, the people around her have come to either discount her emotions or feel sorry for her. Linda's "self-sacrificing," her demanding hopelessness, and her "loud" but unspoken needs evoked intense separation guilt in both Biff and Hap as they grew up and tried to individuate. While Hap has remained deeply enmeshed with Linda, Biff has tried to physically distance himself to avoid his strong separation guilt.

Linda's maladaptive relational pattern is to "deny" her own needs, passively comply with others, and then feel taken advantage of. Her frustration or anger is indirectly expressed via sarcasm, a contemptuous tone, and a hostile demeanor. I am careful not to collude with Linda in denying her needs or allowing them to be expressed indirectly. I am careful not to shame her, but I wish to find and affirm her underlying anger about never having any of her genuine emotional needs met directly in her life. Linda is also adept at taking care of others, will often tell me how "well" she is doing, and how I have helped her. Although these compliments feel good, they also present a way for Linda to deny her own needs in our relationship and are her way of asking me to meet her needs indirectly. To change this problematic relational pattern, I want to help Linda start becoming more connected to her own authentic voice and to be able to own and express it more directly. One way I have tried to attain this goal is to express interest in her concerns as she sees and experiences them. I am also careful to address the areas *she* feels she is "not doing well in" in a way that is nonshaming and normalizes her struggles. I am aware that when Linda feels most helpless she tends to ask me for advice. I then feel pulled to become the "all knowing" therapist and to lead or direct her, knowing full well that I will have to "fail" her as everyone else does.

Aware of my own countertransference propensity here, I monitor my own need to give advice and be "helpful," which I realize would undermine her self-efficacy and confirm her underlying feelings of inferiority and failure and reenact this prototypic relational pattern.

In order to provide Linda with a "corrective emotional experience," the first focus of treatment was to establish a collaborative working alliance. I pursued this goal by helping Linda choose the topics of discussion, trying to enter her subjective worldview, empathizing with her distress, and validating her own feelings, perceptions, and actions whenever I could. It was also important for Linda to find greater safety in our relationship so she could begin exploring her conflicted feelings and her own authentic voice. Because Linda induced so much shame in others, I believed a longer-term goal was a need for her own shame reduction through family-of-origin work. In pursuit of this goal, Linda first needed to be able to recognize and acknowledge to herself when she was experiencing shame in current relationships. She then needed me to help her tolerate these feelings long enough to begin connecting these shame reactions to the original developmental interactions that left her so shame-prone. Linda's denial of her shame in particular, and other conflicted feelings as well, is linked to her lack of ability to experience, identify, or name, and then express them.

Thus, an important aspect of this treatment focus was to help her name her feelings and then role-play with me more effective ways of expressing them. To this end, Linda and I worked on the use of "I messages" (which heighten the sense of responsibility and ownership of feelings and needs). We also began journal writing to further immerse Linda in processing her experiences more directly and fully rather than escaping from these via denial. Linda also needed to become more assertive and less passive-aggressive in her relationships. Said differently, I wanted Linda to learn that she had "a right to her own life" and to develop the ability to advocate on her own behalf and be more direct about her own needs, wishes, and personal limits. To meet these treatment goals I: (1) looked for every opportunity to genuinely acknowledge Linda's abilities and strengths; (2) made overt her tendency to minimize herself or the legitimacy of her own needs whenever possible; (3) modeled direct and honest communication about our relationship and repeatedly invited her to do the same; (4) used self-involving statements to tactfully but directly address her attempts to manipulate me; and (5) reinforced her whenever she honestly and directly expressed a need or feeling by expressing my genuine pleasure in this important new behavior and communicating how I felt closer or more connected to her when she took this risk.

In sum, Linda needed to develop a clearer and fuller sense of self, of who she was as a person, including her disappointments, expectations, needs, wishes, and strengths. Although Linda was comfortable in her role as a wife and mother, she lacked a more fully developed personal identity and connectedness to her true authentic voice. To meet this goal, I also wanted to work with her on expanding her social network and on identifying interests, hobbies, and pleasurable activities. Finally, since Linda had relied first on her father and then on both of her husbands to make decisions, she lacked problem-solving and decision-making skills, which became another focus of treatment.

3. Developmental Context

In Linda's family-of-origin, her father was authoritarian and her mother maintained a submissive and passive role -- the children did not feel valued as individuals. Her mother was ill throughout her life, and Linda was scripted into a parentified care-taking role. Communication was superficial, and genuine feelings were not discussed. Her father had a temper and often became angry -- everyone in the household was intimidated by and resentful of him, and interpersonal conflicts went unaddressed and unresolved. Linda tended to recreate important aspects of these formative relational patterns in both of her marriages by choosing men who presented her with role expectations and

relational problemssimilar to those she had experienced with her
father. Sadly, the trauma of her first husband's suicide and
the shame-evoking need for secrecy surrounding this event caused
her to be even more compliant and resentful in her second
marriage.

As a child, Linda was deprived of secure relational ties
with either her mother or her father. She was responsible for
taking care of her mother, which allowed her to fashion some
sense of connectedness and self, but this also created anxiety
and resentment because she was usually unable to "do it right."
More important, she had no opportunity to be a child and have
her own needs responded to. Her father's temper outbursts and
his demand for unquestioning obedience also contributed to
Linda's resentment regarding not being worthy of attention or
being able to speak up on her own behalf. It was not safe to
acknowledge this, of course, because "having her own mind" would
threaten her already tenuous ties to both of her parents and
further evoke her mother's disapproval and her father's anger.
Saying it most simply, Linda learned to suppress her own
authentic voice in order to survive.

Linda's failure to find secure relational ties in her
family-of-origin has carried over into her current interpersonal
relationships. Her husband and sons deny her needs and take
advantage of her, and she, in turn, denies their true needs and

responds in controlling and invalidating ways. As a child, having an emotional need engendered anxiety and the fear of being shamed in Linda. As an adult, these same emotional reactions are evoked at the thought of asking for anything for herself. Based on her relational templates, Linda still expects her needs to be rejected, as they were by her father (who became punitive, and by her mother, who experienced these as a burden and disconnected or "moved away." These problematic relational schemas have affected her ability either to recognize or act on her needs and to reach out to others. This contributes to Linda's self-imposed isolation from most social encounters. Linda used compliance and indirect communication to cope in her childhood, and she uses the same coping strategies today. Linda's core pathogenic belief is that she must please others or she will be hurt and rejected -- and this compliance fosters much shame and irritable resentment. Further, Linda feels overly responsible for other people and their problems, compelled to "fix" them, and ends up feeling inadequate and "bad" as she inevitably fails to "fix" them. As a result, Linda carries a tremendous amount of guilt over her first husband's suicide. Her childhood left her with the pathogenic belief that she is basically unlovable and only worthwhile when she is taking care of others and meeting their needs. She is thus faced with the dilemma of having to take care of others to be

valued, resenting this demand, and the expectation that she will
be ineffective in her responsiveness, which contributes to her
irritability and depression.

4.   Therapeutic Process

Therapy began with my attempts to establish a collaborative
alliance. I was careful to follow her lead and not press Linda
to disclose anything she didn't want to discuss. I was also
careful to respond to any feelings she revealed, and I validated
her experiences through the use of empathic bids (e.g., "I am
appreciating how difficult it is for you to come here and share
your very personal experiences with me. I respect your courage
to look at your problems and ask for help"). I helped Linda
identify her feelings and thoughts by asking open-ended
questions and encouraging her to focus inward. In the
beginning, Linda was very "other" focused, and I was careful not
to collude with her by encouraging her "storytelling"
tendencies. Gradually, she began to differentiate between where
her responsibilities ended and where others' began. We formed a
collaborative alliance as I let Linda know I realized how
difficult it was for her to disclose personal information and
family secrets. She subsequently acknowledged how embarrassed
and guilty she felt about being disloyal to her family.
However, Linda was visibly relieved to share her feelings and
needs more directly and to find that she was not punished or

abandoned by me in this. Instead, she experienced increased connectedness, which freed her to explore more fully her feelings of being unworthy and the associated shame.

Throughout this process, Linda gave me repeated tests to pass. For example, when she expressed a need to change her appointment time, I willingly accommodated her wish and communicated how happy I was to work with her on establishing a new time. I also emphasized how much I appreciated her directness in asking for what she wanted. As opposed to her past experience, Linda was able to experience being heard and having her needs honored.

During our fourth session, Linda expressed concern that I would see her as "defective" and not really want to be working with her (abandon her). When I told her that nothing she could say or do would make me see her as defective, and that I genuinely enjoyed working with her, she was relieved and began to express painful feelings and grieve over her difficult childhood.

Thus, even though I am aware that Linda can be guilt inducing, I feel compassion for Linda and her life experiences. A turning point in our relationship occurred when I conveyed to Linda how I experienced her "martyr" presentation -- overtly self-sacrificing but laced with hostility and resentment. I told her that it made sense to me why this coping strategy had evolved but that she did not need it with me. I also conveyed how much closer I felt to her when she expressed her needs directly and owned them. By directly and respectfully addressing this, Linda was able to engage with me regarding her needs and associated shame. Her assertiveness increased, and she was later able to transfer this to her relationship with her family. She is a resilient person who survived the pain of being parentified during her childhood and who continued this burdensome role of feeling responsible for taking care of her sons and husband. Upon disclosing this to Linda, she said that nobody had ever understood how difficult it was to have to take care of everyone and how "alone" she had felt. In sharing this with me, Linda was now no longer alone and was being responded to rather than "taking care of." With this corrective experience, Linda began to assess this coping strategy more realistically. As she began to relinquish this role, I observed that she began to act less as a martyr and that her covert anger and demanding attitude diminished.

Occasionally, Linda still recreates problematic relational "taking care of" patterns. She does this by covering up her painful feelings and telling me how much I have helped her. I am aware that this "moving toward" maneuver emerges when she feels particularly ashamed of having needs. This seems to help her deny these needs with me as she has done in her other relationships. She also avoids showing me her anger, fearing I will abandon her. I address this with her respectfully but directly. For example, I frequently check to see if there is something she isn't telling me and, at the end of most sessions, ask her if there was something I said or did that didn't feel helpful or didn't "fit" for her. This invitation has proven to be productive and helps Linda express her concerns that might otherwise go unvoiced.

Linda sometimes puts herself down and minimizes her accomplishments. She often seems surprised when I acknowledge and cheer her efforts, and she is unsure of how to manage this affirmation. For example, she once suggested that since she was doing so well did I think she should terminate our session early? I suggested that this was not necessary and that it was okay to use the session to celebrate her good feelings. As a result, Linda was able to give herself permission to celebrate her accomplishments both in and outside of therapy. On another occasion, Linda complained of our slow progress. I told her I was glad she shared this concern and asked her what we could do to get "unstuck." By doing so, I affirmed her assertiveness and was acknowledging that she had ideas I considered worthwhile. Linda then generalized this stronger stance in the subsequent week by acting more assertive with her friends and directly addressing a conflict.

Linda tends to place me in the role of the all-powerful healer, looks to me to "fix" things and solve her problems, and then nonverbally communicates her sadness and distress because I have failed her. It is important for Linda that I refrain from giving advice, not buy into either role of rescuing or failing her, and continue to take a collaborative stance and foster a shared or working alliance with her.

5. Goals and Interventions

A significant intermediate goal for Linda was to help her foster her own initiative and efficacy. As mentioned earlier, I did this by encouraging her to identify the goals for therapy that she considered important to her own growth. This was new and different for Linda as she was accustomed to meeting others' expectations and not her own. Another intermediate goal was to help Linda increase her self-awareness through identifying her own feelings as well as developing a fuller awareness of the problematic impact she often had on others. This was done by asking her open-ended questions such as: "Linda, what are you feeling right now as I am saying this to you?" or "What do you think your son is feeling as you say that to him?" I also made self-involving statements and process comments: "Linda, as you are telling me of this long and detailed event, I am feeling disconnected from you; almost as if there was a wall between us. What do you think is going on between us? Do you think others have experienced you in this way?" I realized that it was important for me to be consistent with these interventions because it helped Linda get in touch with the shameful feelings that she had been defending against with her storytelling coping strategy. For the first time, she began to recognize the shame and guilt she was evoking in others.

I had several long-term goals for Linda: (1) Linda needed to become more assertive with her family; (2) she needed to begin responding to others in more direct and constructive ways; and (3) she needed to expand her social network. As Linda became increasingly aware of her relational pattern, she realized she could be "heard" by others sometimes and that she "had some rights." I helped Linda learn to express her wants and needs with her family more overtly by having her role-play them with me. Practicing "I Messages" was especially helpful to her.

I further helped Linda explore her needs by addressing her "dreams": "Linda, is there anything that you have always wanted for yourself but that you have put on the 'back burner'?" Linda was able to identify a long-held wish, one of becoming a teacher. This was a secret desire she had "never told to a soul!" Happily, Linda has decided to fulfill her dream by volunteering to teach illiterate adults at her local library two times a week.

As another sign of therapeutic progress, Linda is beginning to advocate for herself at home. Recently, she said she was even considering the possibility of telling her family about her therapy. Due to Linda's "traditional" relationship with her family, we are careful to consider all possible consequences to her new stronger stance. This includes examining her expectations for herself and her family if she informed them of her therapy with me. Ideally, I would like to have provided both couples and family therapy for the Loman's, but realistically, this was not possible. So I chose to work with Linda alone, supporting her in her "existential" choice to find quality within the boundaries of her family situation.

6. Impediments to Change

Several things that I might have done to contribute to Linda's premature termination or unsuccessful treatment were: (1) if I had failed to create a safe environment in which Linda could feel free to talk and explore her problems (e.g., if I became controlling in response to her controllingness, or became withdrawn or defensive in response to her criticisms); (2) if I had not addressed Linda's feelings and needs but rather followed my own agenda (e.g., pushing her to tell her family about her therapy); (3) if I had colluded with Linda in her storytelling method of defending against her anxiety rather than focusing her inward, tracking her anxiety, and helping her explore the childhood wound that contributed to this style; (4) if I had failed to recognize the importance of the stories Linda tells about others and link them to our relationship and what was going on between us; and (5) if I had failed to maintain an interpersonal balance with Linda (e.g., by being either overinvolved and assuming the leadership role she expected and elicited but resented, or underinvolved by withdrawing from her subtle criticism and unspoken disappointment with however I tried to help). As a result of expressing my own authentic voice, Linda felt free to do the same. We were now able to talk about problems that arose between us more directly and to find a new way to relate that did not repeat the familiar problematic patterns.

PART IV: JOURNAL ENTRY

Use the following section to further integrate the material presented, to address unresolved questions or issues, and to further explore your personal reactions.

_____

_____

_____

_____

_____

_____

_____

_____

_____

_____

_____

_____

_____

CHAPTER NINE: AN INTERPERSONAL SOLUTION

PART I: STUDY GUIDE

Part I will help students prepare for course examinations. It consists of three sections: a list of key terms to define; essay questions for central concepts; and sample multiple choice and true/false questions.

Section A: Key Terms

Provide a one or two sentence answer for the following key terms.

ACTIVATION OF RELATIONAL TEMPLATES _____

_____

_____

PATHOGENIC BELIEFS _____

_____

_____

CONTENT LEVEL VS PROCESS DIMENSION _____

_____

_____

AFFECTIVE THEMES _____

_____

_____

CORRECTIVE EMOTIONAL EXPERIENCE_____

_____

_____

CLIENTS' GENERIC CONFLICTS _____
_____
_____

ADDRESSING CURRENT INTERACTIONS _____
_____
_____

INTERNAL FOCUS _____
_____
_____

RESPONDING TO CLIENTS' AFFECT _____
_____
_____

CLARIFYING RELATIONAL PATTERNS _____
_____
_____

VALIDATING CLIENTS' CONCERNS _____
_____
_____

ENTERING CLIENTS' SUBJECTIVE WORLDVIEW _____
_____
_____

THERAPIST RESPONSE VS CLIENTS' EXPECTATIONS _____
_____
_____

Section B: Key Concepts

Consider each of the following key concepts. Be prepared
to write a two or three paragraph essay for each of the
following questions.

1.   Discuss the potential advantages and disadvantages of
     having clients' relational templates activated within the
     therapeutic relationship.

     _____

     _____

     _____

     _____

     _____

     _____

     _____

2.   The interpersonal process must enact a resolution to
     clients' conflicts. What symptoms or signs can therapists
     use to discern when maladaptive relational patterns are
     being reenacted within the therapeutic relationship?
     Discuss both effective and ineffective therapeutic
     responses to these enactments.

     _____

     _____

     _____

     _____

     _____

     _____

Section C: Sample Multiple Choice and True/False Questions

Answer the following sample multiple choice and true/false questions.

1.   When clients' conflicts begin to be replayed in the therapeutic relationship, and the therapist accepts the validity of their concerns and helps the client explore them rather than continue to reenact them:
   a.   the therapist offers the opportunity to negotiate a new type of relationship and resolve the conflict in the current interaction.
   b.   the therapist strengthens the client's reliance on that interpersonal relational pattern.
   c.   the client will become anxious and defensive.
   d.   the therapist encourages the client to reenact the conflict.

2.   When client conflicts are subtly being replayed in the therapeutic relationship, and supervisors or colleagues identify these reenactments, beginning therapists commonly feel:
   a.   inadequate.
   b.   enlivened.
   c.   powerful.
   d.   prepared.

3.   When clients' issues are activated within the therapeutic relationship, the *best* way of responding is for therapists to suggest alternative ways for clients to act with others.

   a.   true
   b.   false

Answers: 1.a, 2.a, 3.b

PART II: SELF-REFLECTIONS

The material presented in the text is often personally evocative and challenging. The questions below are designed to help students process and integrate their reactions to this information.

1. For most beginning therapists, making process comments and addressing conflict in the therapeutic relationship can be anxiety provoking. Consider past experiences you have had with others where forthrightness was used in a helpful way, and those in which it was used insensitively.

_____

_____

_____

_____

_____

_____

_____

_____

_____

_____

_____

_____

_____

_____

2.   Recall a time you felt effective in responding to someone's
     problem and realized you had a positive impact and had
     become important to him or her. Looking back, were you
     able to allow yourself to be this influential, or did you
     feel compelled to retreat from the success and diminish the
     impact on yourself or the other?

_____

_____

_____

_____

_____

_____

_____

_____

_____

_____

_____

_____

_____

_____

_____

_____

3.   How do you evaluate yourself in terms of your ability to
     speak directly or forthrightly about conflicts occurring
     between you and important others? What is the biggest fear
     evoked in you by making process comments or discussing what
     is occurring between the two of you?

_____

_____

_____

_____

_____

_____

_____

_____

_____

_____

_____

_____

_____

_____

_____

_____

4.   What was your original "Dream" about becoming a therapist? How has your Dream regarding counseling changed over time?

_____

_____

_____

_____

_____

_____

_____

_____

_____

_____

_____

_____

_____

_____

_____

_____

_____

_____

PART III: SUPPORTIVE READINGS AND CASE STUDIES

The following Case Study of the Gared family was taken from the movie "Ordinary People", directed by Robert Redford. It illustrates how to use and apply the Case Conceptualization/ Treatment Planning format presented in Appendix B of the text to an actual case. Readers are encouraged to view the movie prior to reading the case study and to attempt to write a Case Conceptualization/Treatment Plan for Conrad.

Section A: Case Conceptualization Guidelines: Case Study Of the Gared Family

1. Formulate the Problem(s)

As the movie opens, we find Conrad Gared suffering from the consequences of a recent boating accident that he had been involved in. Conrad's older brother, Buck, drowned in this accident, although Conrad had tried everything he could to save him. After this event, Conrad unsuccessfully attempts to commit suicide. After being in a hospital for four months and home for a month and a half, things still are not going well for Conrad. He is depressed, uninterested in his daily activities, and does not have much energy. Conrad's symptoms are consistent with a diagnosis of Major Depression. In addition to the symptoms mentioned, Conrad also has recurrent thoughts of suicide and displays feelings of worthlessness and excessive and inappropriate guilt. Conrad also has symptoms consistent with

Posttraumatic Stress Disorder. He has experienced an event that is outside the range of usual human experience by watching his brother drown in the boating accident while trying to help him. He has continuous recollections of the event and at times relives the accident in the form of recurring dreams. Conrad tries to avoid the feelings associated with the event, is generally detached from others, and is limited in the emotions he expresses. Conrad's father, Calvin, is aware of these difficulties and asks him, "Call that doctor yet?...we need to stick to the plan." Conrad responds, "The plan was if I *needed* to call him." Conrad grudgingly decides to enter therapy.

Although Conrad initiates therapy, he is genuinely reluctant to do so. Both having problems and seeking help for them are inconsistent with his interpersonal coping style. Conrad has developed a "moving away" coping strategy. This coping style is characterized by withdrawal from and physical avoidance of others. Conrad has used this style to cope with and defend against problems throughout his life. An individual with this type of coping style has learned that withdrawal and avoidance serve to manage emotional conflicts and interpersonal problems. This coping style failed to manage the overwhelming sadness and guilt evoked by the boating tragedy, however, and precipitated his suicide attempt. As we will see below, Conrad has not been able to reconstitute his coping style and remains depressed.

2. Developmental Context

Conrad acquired this maladaptive "moving away" relational pattern early on in his life because he never achieved a secure attachment to either of his parents. As a child, his age-appropriate needs for warmth, affection, and love were left unmet--probably as a result of his parents having marital conflicts and unresolved developmental issues of their own. In particular, it seems that in an unconscious attempt to distance or control her own conflicted identity issues, Beth, Conrad's mother, has "split" the good and bad aspects of herself and projected them onto her children. Buck was the child she selected to carry the "good" aspects of herself, and he became her favorite. He was considered the smartest, the funniest, the strongest, etc. As a result, he acted as scripted. This can be seen in the scene where Conrad sees his mother laugh and then remembers an occasion where his mother is laughing at Buck's antics. She is completely animated, and Buck is center-stage. Poignantly, rejected and ignored, Conrad is distantly watching from the sidelines.

It is also interesting to note Beth's interaction with Buck. It is almost as if she is being flirtatious with him by the way she is reclining, tossing her head back and laughing,

and throwing things at him. It is almost like a scene you would
see between a girlfriend and boyfriend and is significant in
that it may be a good characterization of Beth and Buck's
relationship overall. Certain emotional needs are being met for
Beth through Buck that should be met instead with her husband--a
primary, cross-generational mother-child alliance had been
formed between Beth and Buck.

In contrast, Beth seems to have projected the "bad" or
unwanted aspects of herself onto Conrad. She treats Conrad as a
scapegoat as she blames many problems on him--including covert
blame for surviving Buck. Further, there is reference to Beth's
reducing the significance of Conrad's attempted suicide to an
inconvenience--for example, not being able to get the blood
stains out of the towels! Conrad, already scripted as the "bad"
or unwanted son, is pained and "buys" into this. In one of his
therapy sessions, for example, he states that he can't blame her
for being upset with him, it is "impossible after all the shit I
pulled" (referring to the blood on the towels and in the grout
on the floor, etc.). Sadly, he cannot even validate the
seriousness of his own traumatic losses and familial problems.
Throughout the movie, Beth is emotionally cold and withdrawn
toward Conrad. She does take care of his basic survival needs,
but not much more. It seems she is most connected to Conrad

when he fulfills the triangle by enabling the alliance between her and Buck.

Unable to find a buffer or other relational "out" from his dilemma with his mother, Conrad was not able to form a secure relational attachment to his father either. Instead, Conrad seems to have taken on the responsibility, in part, for emotionally taking care of his father. Since Beth turned her attentions toward Buck, Calvin may have leaned toward Conrad for some level of emotional fulfillment. Little Conrad, desperate for any connection with one of his parents, probably fell easily into this secondary parentified role, the theme of which continues throughout the movie. For example, when asked by Dr. Burger in his first therapy session why he decided to come to therapy, Conrad answers, "So people can quit worrying about me." Dr. Burger asks, "Who's worried about you?" Conrad replies, "My dad, I guess...my father mostly. This is his idea." This is also exemplified when he quits the swim team. He tells Dr. Burger that he doesn't want to tell his dad because, "...it would kill him."

Not being able to form secure attachments with his parents would have created an intense anxiety in little Conrad, and thus, his generic conflict would have been created. In order to ward off this anxiety and rise above the conflict, he had to develop a compromise solution that would prevent him from

experiencing shame associated with his unmet needs. Thus, he
began blocking his needs in the same way they were blocked in
the environment, deeming them as unimportant and insignificant
relative to the needs of others in his family. Invalidating
himself in this way would allow him to make sense of his
environment by allowing him to continue to view his parents'
neglectful actions toward him as sensible. This loss of self or
the authenticity of his own experience, in turn, would provide
him with some control over his feelings instead of experiencing
inefficacy and suffering the shame of being rejected. In this
way, he developed the pathogenic belief that he was not lovable,
felt critical or contemptuous of himself for having the need to
be cared about and wanted, and coped by distancing himself from
these feelings and needs. By turning against himself in these
ways, he was acting toward himself in the same way his mother
acted toward him--the first part of the compromise solution.

The second part of this compromise solution is shown by the
way Conrad blocks his anxiety-arousing needs by reenacting in
current relationships the same conflict he experienced with his
mother. Conrad is emotionally detached from friends and others
in the same manner his mother is emotionally detached from him.
Also, it appears that he chooses or selects to associate with
people who are emotionally distancing and removed themselves.
Finally, Conrad elicits the same unsatisfying response from

others in his relationships as he had from his mother. It should be noted that Conrad's being distanced from and by friends may be due to Buck's death and his attempted suicide and may be reality-based reactions. Understandably, people do not know how to act around someone who is seriously depressed, just released from a mental hospital for slashing his wrists, and had his brother drown. In a negatively spiraling interaction chain, he becomes further self-conscious and uncomfortable with this hesitancy in others. But this maladaptive relational pattern goes beyond these situational crises. We can't understand Conrad without appreciating that even before Buck's death this maladaptive relational pattern would have existed in order to keep the pain of his primary maternal rejection at bay. If he was confronted with a relationship in which someone found him loveable and attempted to be unconditionally responsive and warm, it would have aroused the pain of the original deprivation Conrad experienced throughout childhood and adolescence.

Even though Conrad would be defending against the pain of his unmet needs, he would also be simultaneously trying to "rise above" his unmet needs and to have them met indirectly. Conrad tried to rise above his generic conflict by adopting a "moving away" interpersonal coping style. This inflexible interpersonal coping style is the other side of the generic conflict and compromise solution. For example, Conrad learned to avoid seeking direct contact (i.e., interest, affection, support, or love from his mother and father) and, instead, became emotionally self-sufficient and withdrawn from his mother and responsible for the happiness of his father. With his mother, the only emotional connection he could fashion was through Buck. This is shown in the earlier-mentioned scene where Conrad looked on as Buck entertained their mother. In this way, he could have some kind of relational tie with his mother, for she was pleased with him when he stepped aside and let Buck be in the forefront of any situation. With his father, he maintained a connection to him when he did something his father wanted him to do (i.e., go to therapy, be on the swim team, achieve, be happy, and not have any problems, etc.).

It is also interesting to observe Conrad's many perfectionistic behaviors. At school he got straight A's. He was not just on the swim team--he was the *best* swimmer on the team. At home, he was "hard on himself." In fact, it is near the end of the movie when his father, who has begun to change as a result of his own therapy, points this out. Calvin begins to yell at Conrad for attempting to blame himself for his mother's angry departure for an extended trip alone. He then catches himself and apologizes. Conrad tells him it is all right to yell at him the way he used to yell at Buck. Calvin just replies, "Buck needed it. You were always so hard on yourself,

I didn't have the heart." These perfectionistic behaviors also represent the part of Conrad that wants to be noticed and loved. In summary, this struggle of rejecting himself and others, yet indirectly seeking connectedness and being valued for his excellent behavior, is Conrad's compromise solution to his generic conflict.

Further, this defensive, moving away coping style would have been turned into a "virtue" by Conrad. He would feel that he does not need love and recognition; he is self-sufficient and this makes him stronger and "better" than others who need to be loved. From this would come his entire sense of self-esteem and sense of self. He *is* his ability to remain aloof and self-sufficient, so he can stand back to let his brother's needs be met first. In actuality, however, this sense of self would only be a contrived and rigid substitute for the genuine sense of self he lacks because he cannot rise above the original conflict or unmet needs. As a result, Conrad's coping strategy is fragile, and when it fails, his entire sense of self is threatened--resulting in the severe consequences of a major depression and a serious suicide attempt.

Buck's death causes a failure in Conrad's interpersonal coping strategy. According to his maladaptive relational pattern, the only way he could ever be connected to his parents was by letting his needs come last. In particular, to be connected with his mother, he would have to let Buck have

center-stage. Consequently, he believes he should have been the

one that died, not Buck. So in addition to the reality-based

response of guilt for not being able to save Buck and pain for

the loss of a loved one, Conrad believes he has betrayed his

mother by his needs coming first and his "letting" Buck die.

This situation profoundly threatens his already fragile

relational ties to her and brings the shame-based "reality" of

being a hated child to the surface. To repeat, his entire sense

of self is severely threatened, and his shame-based sense of

self (which was kept hidden by his interpersonal coping

strategy) has been revealed. This arouses the original guilt,

helplessness, and shame that accompanied his unmet needs for a

secure relational tie as a child. His suicide attempt results

from the excruciating pain this causes him, as well as the

terror of consciously experiencing his mother's rejection,

contempt, and/or hate rather than splitting away from it as he

had in the past.

Because his suicide attempt is unsuccessful, he still has

the excruciating pain of his generic conflict left to deal with,

and he attempts to use his moving away coping strategy to

distance himself from these painful emotions. This is

impossible, however. Conrad returns home from the hospital to a

disrupted family system whose homeostatic relational patterns have become completely destabilized. Buck is no longer there to

be Beth's primary ally. It is obvious that her resentment of

Conrad is not just because he lived, he also took away the

person that gave her emotional fulfillment, as well as her

"good" self--the one she projected onto Buck. Beth expresses

this resentment toward Conrad with constant guilt-inducing

reminders, albeit passive-aggressive, of her resentment. This

is exemplified at the breakfast table one morning when Conrad

does not want to eat the cinnamon toast his mother made for him.

He says he is not hungry, but she overreacts by grabbing his

plate and throwing it down the garbage disposal in a very

martyrish way. Furthermore, Beth actively works to cut

relations with him. This is seen when Conrad comes home from

school one day and hesitantly tries to initiate a conversation

with Beth. Uncharacteristically, she begins to become engaged

in the conversation. Conrad is talking about having to work on

geometry and becomes more animated as she is responding. She

says she always had a hard time with geometry. Then, suddenly,

it is as if she catches herself, "Geometry?" she asks herself,

"I don't think I ever took geometry..." The anxiety in Conrad

is palpable as she abruptly cuts off from him and goes into her

room, shutting the door behind her. It is obvious that all

relational ties with his mother are gone, no matter how tenuous

they were before Buck's death. Conrad is consumed by guilt for hurting his mother and anguished by painful separation anxieties--he cannot reestablish the ties his mother has actively cut.

It is interesting to note that at this point in the family constellation Beth begins to establish a new alliance with Calvin. A primary parental coalition is normal and healthy, but Beth appears to be trying to alienate Calvin from his son. For example, when Conrad blows up at his mother and then leaves the room while he and his dad are decorating the Christmas tree, Beth tries to talk Calvin into not consoling him, "Don't indulge him," she states, completely invalidating the seriousness of Conrad's emotional state. Another example occurs when she tries to talk Calvin into going away for Christmas. Calvin is uneasy with the suggestion since Conrad is in therapy and is struggling to recover. Beth has no regard for Conrad's needs at all. These situations exemplify an attempt by Beth to force Calvin to choose between fulfilling her needs or fulfilling Conrad's. Additionally, her doing this is triangulating Calvin into the conflict between her and Conrad. Buck is no longer there to be a buffer, so the familial problems that existed before Buck's death have now been amplified.

3. **Treatment Focus and Therapeutic Process**

It is at this point that Conrad enters therapy. Dr. Burger begins by treating Conrad in a businesslike manner. He asks

information-gathering questions in a very matter-of-fact way,
and it does not seem like he is successfully establishing a
collaborative alliance with a client whose brother has just died
in a boating accident and who has tried to commit suicide. Dr.
Burger nonchalantly refers to Conrad's brother dying,
"...boating accident, was it?" and, "Wanna tell me about it?" as
if he is completely distanced and removed from the incident, as
well as removed from Conrad. This reenacts Conrad's generic
conflict. Conrad is familiar with relating to people in a
distancing and self-controlled way and expects people to react
to him in such a way. He does not believe he is lovable or that
he is worth getting close to or being responded to with care or
warmth. Consequently, he proceeds to react to Dr. Burger in his
moving away style and does not respond to Dr. Burger's question
regarding his brother's death.

So it is not surprising that as Dr. Burger continues with
the questions, Conrad states that already he does not like what
is happening and is not happy to be in therapy (ironically, this
is a bold and important process comment, albeit resistant, but
Dr. Burger does not explore it). The resistance and distancing
maneuvers continue to be enacted during this first session as
Conrad gives Dr. Burger a hard time about what times he is
available for sessions. These are examples of transference
tests. Conrad would like to be responded to in a loving way and

is (unconsciously) trying to ascertain whether he will be
responded to in a different manner than others in his past or if
Dr. Burger will become angered or distanced. If he does become
angry, he will fail the test because he will be responding in
the same unsatisfying way as others in his life have. This is
Conrad's expectation or relational template. Conversely, if Dr.
Burger does not distance from Conrad, Conrad would probably
still remain distant. It is his interpersonal coping style to
remain distant and act self-sufficient in order to avoid the
anxiety the situation provokes in him, as well as to continue to
try to elicit a distancing response in the therapist. At the
same time, if Dr. Burger consistently passes these transference
tests and provides a more corrective or reparative relationship,
Conrad would begin to feel safe about moving into exploring more
threatening issues. What does happen is that Dr. Burger does
not get angry or distanced and therefore passes the test. This
begins the gradual process of the disconfirmation of his
pathogenic beliefs and expansion of his core relational
patterns.

Conrad's distancing style is illustrated to some extent in
most of the following sessions. For example, in the second
session, when Dr. Burger is pushing to find out what is making
Conrad so jumpy, Conrad just avoids the topic by challenging Dr.
Burger (in a "moving against" manner): "Maybe I need a

tranquilizer." When Dr. Burger does not get intimidated by
this, Conrad starts giving him a hard time about the clock on
his desk, "I see, you get to tell the time but I can't?" Again,
Conrad is testing Dr. Burger. Conrad is continuing to try to
find out if he could effectively distance Dr. Burger as he has
done with others in his past. At this point, it would have been
easy for Conrad to have elicited a countertransference reaction
in Dr. Burger. If he would have reacted with anger or by
withdrawing, it would have confirmed Conrad's pathogenic belief
that he does not deserve understanding or connection, for here
again someone is acting emotionally distant from him. Dr.
Burger seems to pass these tests, however, for he is not angered
or distanced. This is one of Dr. Burger's strongest points. He
is able to tolerate Conrad's distancing maneuvers and
challenging actions without becoming undone, angry, or
withdrawn.

After these beginning sessions Dr. Burger seems to develop
an integrating focus for the material Conrad was presenting. In
these beginning sessions, the recurrent theme seems to be
Conrad's "lack" of emotions or, perhaps more accurately, his
tendency to distance from his emotions. Conrad talks about how
John Boy Walton would have said something about the way he felt
if his brother had died, but Conrad could not because he did not
feel anything. Also, later, he fights getting angry at Dr.

Burger because it takes "too much energy." This is in keeping

with Conrad's generic conflict of remaining detached from his

emotions. Dr. Burger seems to be aware of Conrad's primary

presenting affect of being aloof, "lacking" in emotional

expression.

Dr. Burger may have also developed the tentative hypothesis

that Conrad is trying to elicit distancing behavior in him, is

threatened by being overwhelmed by emotion, and that resistance

will be (and is being) expressed by continuing to perform

distancing behaviors. Even without knowing Conrad's family

background, the therapist could formulate the tentative working

hypothesis that Conrad must have had a caretaker who treated him

in an emotionally distancing manner, and as a result, he treats

himself and others in this manner. Additionally, he could

postulate that there is a part of Conrad that wants to be

emotionally close to others, but the pain and accompanying shame

of this deprivation is intolerable. This is an old wound.

Putting that together with the multiple stressors Conrad has

experienced (Buck's death, his attempted suicide, Buck's absence

from the family constellation) helps him understand Conrad's

problems and provide a treatment focus.

4. Goals and Interventions

Indeed, Dr. Burger seems to be aware that Conrad does not

experience his feelings. He makes Conrad's experiencing his feelings a treatment goal. During the fourth session, Conrad admits that he is not going to feel anything, but Dr. Burger does not let him off so easily and (provocatively and intrusively) goads him into anger. Following Dr. Burger's questionable interventions, Conrad almost explodes. He holds himself back, however, because it "takes too much energy to get mad," and states that when he "feels, all he feels is lousy." Again, Dr. Burger does not accept this and finally pushes hard enough to make Conrad explode. Conrad is amazed to find that he is still alive and well after erupting with his anger. This is due to Dr. Burger's more effective intervention of approaching his expressed emotions caringly and providing a secure holding environment for them. In effect, this intervention accomplished the goal of showing Conrad that neither he nor Conrad were overwhelmed by the emotions and that therapy was a safe place for him to express them. This is the beginning of Conrad's emotionally reexperiencing his original unmet need and a step toward beginning to integrate the feelings that accompany it.

Although Dr. Burger's inciting anger in Conrad seems effective in this case, a better intervention for approaching Conrad's lack of feeling may have been to ask Conrad to explore what the danger might be in experiencing emotions. In this way, the therapist would not be bidding for the content but for the process. The resistance (e.g., Conrad might have said such

things as "I'm afraid you'll make fun of me" or "I'm afraid of
going out of control and I'll never stop crying") could have
been identified in this way, worked through together, and Conrad
would have become further engaged in the treatment process.

Regardless, Conrad's generic conflict starts to emerge when
he makes the connection that it is he who does not forgive
himself for surviving Buck, just as his mother does not forgive
him. It is as if on some level he realizes that she chose Buck
over him and that has caused him pain. This makes it important
that in therapy the therapist will act in a manner that is
different and healthier than the client has experienced with
others. This appears to be another one of Dr. Burger's goals:
to provide Conrad with a corrective emotional experience.
However, at this point Conrad still does not fully realize that
he has acted toward himself in the same way his mother acted
toward him. He has taken on her attitude toward him, and he
still does not believe that he is lovable. This becomes obvious
when he invalidates his dad's love for him by saying that his
father may love him but that he loves everybody. Conrad cannot
admit that he is loved because he is worth loving. This leads
Dr. Burger to point out to Conrad later that, "There is someone
else you have to forgive." This is difficult for Conrad to
fathom, especially because in comprehending it he would have to
experience the feelings associated with the original

deprivation, i.e., his generic conflict. In particular, responding to himself with love and forgiveness would open the old wound of not having been responded to by his parents with emotional availability and responsiveness as a child. He must experience the emotions connected to his old wound and accept both the good news--that he indeed is lovable, and the bad news--that his mother has rejected him. Additionally, he must forgive his parents for their shortcomings in order for him to forgive himself.

By responding differently than others have in the past and how the client has come to expect others will respond, the therapist creates the interpersonal safety the client needs to risk sharing or re-experiencing conflicted feelings. The therapist, in turn, helps the client amplify, further clarify, and integrate the feelings by providing a corrective emotional experience--that is, responding in more accepting and affirming ways than the client has received from others. In the final climactic therapy session, Conrad telephones Dr. Burger as he enters an emotional crisis when he learns that his friend Karen (who he had been in the hospital with) had successfully committed suicide. He had been able to stay distanced from his emotions until this point, but Karen's suicide (and possibly Dr. Burger's work) triggered an avalanche of emotion. He wants to not feel it but, "...it just keeps coming!" he cries.

Dr. Burger seems to realize that the greatest point of
change occurs at the moment the client experiences the full
emotional impact of his/her problems. Effectively, he is right
there with Conrad as he is experiencing his emotions. He tells
Conrad not to stop his emotions. Conrad starts repeating, "I've
gotta get off the hook...for what I did." Conrad's generic
conflict starts to emerge as he begins to relive the boating
accident. Dr. Burger does not back off as Conrad's emotions
overtake him. He asks, "What did you do?" Dr. Burger remains
right there with him, responding to each successive feeling in
his affective constellation of emotions, providing a holding
environment and a "real life" experience by taking on the role
of Buck. When Conrad is crying, "Bucky, I didn't mean it!" Dr.
Burger responds, "I know, it wasn't your fault!" Conrad is
experiencing excessive and inappropriate guilt (the first
emotion in his triad). As Conrad continues to blame himself, it
begins to turn into anger (the second emotion in the triad) at
Buck for putting him in charge of the boat. In effect, Buck had
placed all the responsibility (including that of his life) on
Conrad, and Conrad is damn angry about it. Dr. Burger stays
with him, "...and that wasn't fair, was it?" This almost seems
to surprise Conrad--someone is seeing it from his perspective
and validating that it was not fair to him. Nevertheless,
Conrad's anger continues mounting because Buck had told him to

hold on, but then Buck himself lets go. Conrad is furious that
he let go and cannot comprehend why. He always perceived Buck
to be the leader, the strong one. It just does not fit into
Conrad's schema of life and his identity that he might have been
stronger and that was why he lived. Also, being forced into
this role is so alien for Conrad that it threatens life as he
knows it. In essence, it breaks all relational ties with his
mother, which evokes the third emotion in the triad. Dr. Burger
effectively validates this sadness and lets him experience this
while remaining connected to him. Overall, Dr. Burger has
provided a holding environment for each of Conrad's emotions,
which is a powerful and resolving corrective emotional
experience for Conrad.

Following Dr. Burger's validation, Conrad realizes that he
blames himself for being the one that held on. According to
Conrad's coping strategy--and the role he played in his family--
he should have been the one to die, not Buck. He was second to
Buck even though a part of him craved to be first.
Nevertheless, when this need to be first was met, Conrad was
horrified, for this would mean that he would not be loved by his
mother. His mother's "love" was available only when he was
second to Buck or in some way connected to Buck. Thus (in
Buck's absence) to live makes him unlovable. Further, realizing
the pain of his mother's rejection, Conrad becomes intensely

anxious and says he is scared to live. However, Dr. Burger

says, "You're here, and you're alive...it is good, believe me."

Conrad asks, "How do you know?" Dr. Burger states, "Because I'm

your friend." He is, in effect, saying that he loves Conrad

even though he lived. He is lovable. This provides the

corrective emotional experience necessary for Conrad to begin

resolving his profound conflict. Dr. Burger has established a

secure emotional tie with Conrad to replace the broken tie

between Conrad and his parents. Conrad can now begin to live

and be emotionally alive, rather than remain depressed and

living as dead.

5.   Impediments to Change

If Dr. Burger had remained detached or disconnected from

Conrad, as he appeared to be in the first session or two,

therapy would have failed. Or, if Dr. Burger had started out as

being connected to Conrad, but had become distant as Conrad

began to experience emotions, he would not have been able to

help Conrad. Conrad would have remained withdrawn from Dr.

Burger and cut off from his own feelings. Both of these

situations would have reenacted his old relational patterns, and

Conrad would not have been safe enough to risk reexperiencing

and sharing his emotions. It would have further reenacted and

confirmed Conrad's pathogenic beliefs about himself and others,

and intensified his coping strategy of remaining detached from others and of having others remain detached from him. Conrad probably would have dropped out of therapy and become further cut off from others, alienated from himself and his genuine emotions, and more depressed and self-destructive if this type of problematic interaction continued to occur in therapy.

The movie ends shortly after Conrad experiences the full impact of his emotions, but in real life therapy would continue. Conrad would have to work through what had happened and come to terms more fully with the grim and profoundly unwanted realization of being a hated child. Although his mother did not love him as she loved Buck, he would need to further integrate the realization that he is lovable and no longer needs to remain emotionally distant from others. Also, he would need to continue working on not feeling responsible for his father's happiness and connect with him in a different way. As the movie suggests at the end, Conrad would now probably have the power to do this due to his experiencing, expressing, and containing each feeling in the affective triad that was related to both the boating accident and, more important, to his life-long generic conflict. Dr. Burger would continue to assist Conrad in accepting that he cannot change the past and that he may or may not be able to have the type of relationships with his parents in the future that he would want. However, ultimately, Conrad could learn that he will be able to have healthy and fulfilling

relationships with others in his future. One child is
tragically lost, but another now has a chance to live.

PART IV: JOURNAL ENTRY

Use the following section to further integrate the material presented, to address unresolved questions or issues, and to further explore your personal reactions.

_____

_____

_____

_____

_____

_____

_____

_____

_____

_____

_____

_____

CHAPTER TEN: WORKING THROUGH AND TERMINATION

PART I: STUDY GUIDE

Part I will help students prepare for course examinations. It consists of three sections: a list of key terms to define; essay questions for central concepts; and sample multiple choice and true/false questions.

Section A: Key Terms

Provide a one or two sentence answer for the following key terms.

WORKING THROUGH _____

_____

_____

AFFIRMING NEW RELATIONAL PATTERNS_____

_____

_____

GENERALIZING CHANGES _____

_____

_____

COGNITIVE AND BEHAVIORAL REHEARSALS_____

_____

_____

SELF-DISCLOSURE_____

_____

_____

ANTICIPATING WHEN OLD COPING STRATEGIES ARE LIKELY TO BE EVOKED

_____

_____

_____

CRITICAL INCIDENTS _____

_____

_____

REAFFIRMING CLIENTS' ABILITIES TO CHANGE_____

_____

_____

FAMILY-OF-ORIGIN WORK _____

_____

_____

COUNTERTRANSFERENCE _____

_____

_____

BLAMING CLIENTS' PARENTS _____

_____

_____

GRIEF WORK _____

_____

_____

THE "DREAM" _____

_____

_____

TERMINATION SEQUENCE _____

_____

_____

NATURAL ENDINGS _____

_____

_____

UNNATURAL ENDINGS _____

_____

_____

THERAPEUTIC VS PROBLEMATIC ENDINGS _____

_____

_____

Section B: Key Concepts

   Consider each of the following key concepts. Be prepared
to write a two or three paragraph essay for each of the
following questions.

1.   Describe the sequence of steps that often occurs during the
     working through process. What are the necessary tasks for
     the therapist, and what are the necessary tasks for the
     client?

_____

_____

_____

_____

_____

_____

_____

2.   How does family-of-origin work facilitate improvement in
     clients' current interpersonal relationships?

_____

_____

_____

_____

_____

_____

_____

3.   Discuss ways therapists can facilitate or impede clients'
     family-of-origin work during the working through process.

_____

_____

_____

_____

_____

_____

_____

_____

_____

4.   Explain why it is important for therapists to help clients
     articulate their "Dreams." How can therapists help
     clients' find their own voices and better formulate
     personal visions for their own lives?

_____

_____

_____

_____

_____

_____

_____

_____

5. Why is it important for therapists to address termination
    forthrightly? Discuss both effective and ineffective ways
    to address therapeutic endings.

    _____

    _____

    _____

    _____

    _____

    _____

    _____

    _____

    _____

6. Discuss how your unnatural ending with clients can be
    addressed effectively and ineffectively.

    _____

    _____

    _____

    _____

    _____

    _____

    _____

Section C: Sample Multiple Choice and True/False Questions

Answer the following sample multiple choice and true/false questions.

1.   Using cognitive behavioral interventions and providing information, suggestions, and guidelines to help clients generalize changes with the therapist to others:
   a.   is the only focus of the working through process.
   b.   can be very helpful in the working through stage.
   c.   is always better utilized in the initial stages of therapy.
   d.   is ineffective in the working through process.

2.   The therapist's ability to address termination forthrightly will:
   a.   leave clients feeling powerless and out of control.
   b.   reenact clients' past conflicted feelings over termination.
   c.   give clients a mastery experience by allowing them to become active, informed participants in the termination.
   d.   not benefit clients--openly addressing termination is rarely useful.

3.   When therapists and clients covertly agree to avoid their impending separation, clients' past problematic endings with others are often metaphorically reenacted with the therapist.
   a.   true
   b.   false

Answers: 1.b, 2.c, 3.a

PART II: SELF-REFLECTIONS

The material presented in the text is often personally evocative and challenging. The questions below are designed to help students process and integrate their reactions to this information.

1. Coming to terms with your own developmental history involves the exploration of the "good news" and the "bad news" of your familial relations. Describe some of the facilitative influences that have helped you achieve the successes of the present, and some of the impeding influences that have diminished your capacity for sustained intimacy and your ability to take pleasure in your successes.

_____

_____

_____

_____

_____

_____

_____

_____

_____

_____

_____

_____

_____

_____

2.   Our personal Dreams enliven us and infuse meaning into our
     lives. Describe your Dream. How can you differentiate it
     as an expression of your authentic self and genuine
     interests, as opposed to childhood strategies of coping
     with unmet developmental needs?

_____

_____

_____

_____

_____

_____

_____

_____

_____

_____

_____

_____

_____

_____

3.   Recall important endings you have experienced both as a
      child and now as an adult. Can you identify unifying
      patterns in the ways you felt and responded? How might
      your propensities to respond to endings be problematic with
      clients during the termination phase of therapy?

_____

_____

_____

_____

_____

_____

_____

_____

_____

_____

_____

_____

_____

_____

_____

_____

_____

4. In Chapter One we discussed how the therapist needs to:
    (1) be able to establish and maintain the therapeutic
    alliance, (2) conceptualize clients' dynamics and provide a
    focus for treatment, and (3) enter clients' subjective
    worldview and respond to their feelings. At this point in
    your clinical training, how do you evaluate your ability in
    each of these three arenas? Refer back to Chapter One
    (p.7) and see how your assessment of yourself has changed
    on each of these dimensions at this later stage in your
    development.

_____

_____

_____

_____

_____

_____

_____

_____

_____

_____

_____

_____

_____

PART III: SUPPORTIVE READINGS AND CASE STUDIES

After reading Chapter Four, you completed the self-assessment scale for interpersonal process skills. Now, with further experience and understanding behind you, assess yourself again on these essential clinical skills. Which dimensions have you begun to incorporate or make your own and which remain challenging for you or personally incongruent?

Section A: Clarifying Therapist's Interpersonal Style

1.   Therapist makes process comments or finds other ways to openly discuss distortions, misunderstandings, and other potential problems that may be occurring between the client and the therapist.

       1   2   3   4   5   6   7
not at all   characteristic   extremely
characteristic   characteristic

2.   Therapist actively tries to understand the client's socio-cultural context and how race, religion and gender have shaped *her** subjective worldview.

       1   2   3   4   5   6   7
not at all   characteristic   extremely
characteristic   characteristic

3.   Therapist accurately identifies and reflects the central meaning or emotional message in what the client has just relayed.

       1   2   3   4   5   6   7
not at all   characteristic   extremely
characteristic   characteristic

4.   Therapist helps the client relay her narrative and express thoughts and feelings.

       1   2   3   4   5   6   7
not at all   characteristic   extremely
characteristic   characteristic

Throughout the questionnaire the female pronoun is used to refer to the client.

5.   *Therapist helps the client explore and discuss her personal reactions toward the therapist.*

       1   2   3   4   5   6   7
       not at all   characteristic   extremely
       characteristic   characteristic

6.   Therapist has difficulty attending to how the client may be interacting with the therapist in the same problematic ways that she describes doing with others.

       1   2   3   4   5   6   7
       not at all   characteristic   extremely
       characteristic   characteristic

7.   Therapist has difficulty following the client's lead and staying close to the problems and issues that the client reports as relevant or significant in her life right now.

       1   2   3   4   5   6   7
       not at all   characteristic   extremely
       characteristic   characteristic

8.   Therapist explores developmental events as they arise in the conversation naturally, rather than leading the client back to historical connections.

       1   2   3   4   5   6   7
       not at all   characteristic   extremely
       characteristic   characteristic

9.   Therapist helps the client focus inward on her own thoughts and feelings.

       1   2   3   4   5   6   7
       not at all   characteristic   extremely
       characteristic   characteristic

10.   Therapist has difficulty inviting the client to express whatever feelings she may be experiencing as they occur in the session.

       1   2   3   4   5   6   7
       not at all   characteristic   extremely
       characteristic   characteristic

11.   Therapist encourages the client to explore feelings and thoughts about their current interaction and what is happening in the therapeutic relationship.

       1   2   3   4   5   6   7
       not at all   characteristic   extremely
       characteristic   characteristic

12. Therapist is able to extend herself and actively reach out
     when necessary to maintain the client's engagement in a
     collaborative relationship.
          1   2   3   4   5   6   7
     not at all   characteristic   extremely
     characteristic   characteristic

13. Therapist is nonjudgmental and responds to the client in
     an accepting and understanding manner.
          1   2   3   4   5   6   7
     not at all   characteristic   extremely
     characteristic   characteristic

14. Therapist attempts to conceptualize a treatment focus by
     formulating a working hypothesis about the maladaptive
     relational patterns and interpersonal themes that reoccur
     in the client's life.
          1   2   3   4   5   6   7
     not at all   characteristic   extremely
     characteristic   characteristic

15. Therapist identifies and explores problematic relational
     patterns that might constitute a generic conflict or
     organizing theme in the client's interpersonal
     relationships.
          1   2   3   4   5   6   7
     not at all   characteristic   extremely
     characteristic   characteristic

16. Therapist attempts to link recurrent patterns of behavior
     between the client and others to the interaction between
     the client and therapist.
          1   2   3   4   5   6   7
     not at all   characteristic   extremely
     characteristic   characteristic

17. Therapist looks for unifying themes or patterns that may
     link events that initially appear to be unrelated.
          1   2   3   4   5   6   7
     not at all   characteristic   extremely
     characteristic   characteristic

18. When the client appears to become defensive or resistant,
     therapist helps the client explore what the danger or
     threat is that may have just been evoked.

     1   2   3   4   5   6   7
     not at all   characteristic   extremely
     characteristic   characteristic

19. Therapist is reluctant to focus the client away from
     complaining about or describing the problematic behavior of
     others and toward the client's own personal reactions.

     1   2   3   4   5   6   7
     not at all   characteristic   extremely
     characteristic   characteristic

20. Therapist responds to the client's global descriptions or
     generalized statements about themselves and others by
     seeking further specificity or concrete illustrations.

     1   2   3   4   5   6   7
     not at all   characteristic   extremely
     characteristic   characteristic

21. Therapist is empathic and tries to understand the personal
     or unique meanings of the client's experience from the
     client's subjective worldview.

     1   2   3   4   5   6   7
     not at all   characteristic   extremely
     characteristic   characteristic

22. Therapist is emotionally available and conveys "presence"
     as the client relays her narratives.

     1   2   3   4   5   6   7
     not at all   characteristic   extremely
     characteristic   characteristic

23. Therapist cannot be a "participant/observer" who is
     simultaneously empathic and objective.

     1   2   3   4   5   6   7
     not at all   characteristic   extremely
     characteristic   characteristic

24. Therapist invites the client to discuss what she is
     thinking about the therapist, what she is thinking about
     something the therapist has done, or what she is thinking
     the therapist might be feeling.

     1   2   3   4   5   6   7
     not at all   characteristic   extremely
     characteristic   characteristic

25.   Therapist creates <u>immediacy</u> by using self-involving
      statements or sharing her own reactions to the client.
      1   2   3   4   5   6   7
      not at all   characteristic   extremely
      characteristic   characteristic

26.   Therapist encourages the client to be an active, equal
      partner in understanding problems and initiating changes.
      1   2   3   4   5   6   7
      not at all   characteristic   extremely
      characteristic   characteristic

27.   Therapist helps the client discern discrepancies between
      her public persona or her social roles <u>and</u> her own
      authentic voice and genuine experience.
      1   2   3   4   5   6   7
      not at all   characteristic   extremely
      characteristic   characteristic

28.   Therapist considers unwanted ways that significant others
      have responded to the client in the past and uses this
      awareness to provide a new or reparative response to the
      client.
      1   2   3   4   5   6   7
      not at all   characteristic   extremely
      characteristic   characteristic

29.   Therapist has difficulty formulating working hypotheses
      about how the client's relational patterns could interact
      with the therapist's own personal issues to impede
      treatment.
      1   2   3   4   5   6   7
      not at all   characteristic   extremely
      characteristic   characteristic

30.   Therapist helps the client appreciate how her defensive
      style was originally necessary and adaptive but is no
      longer needed in many current relationships.
      1   2   3   4   5   6   7
      not at all   characteristic   extremely
      characteristic   characteristic

31. Therapist demonstrates the cognitive flexibility and wide
     emotional range necessary to respond to the client's
     varying needs.

        1   2   3   4   5   6   7
     not at all   characteristic   extremely
     characteristic   characteristic

32. Therapist evaluates the effectiveness of her interventions
     by systematically evaluating the client's reactions to
     them.

        1   2   3   4   5   6   7
     not at all   characteristic   extremely
     characteristic   characteristic

33. Therapist feels she can be authentic with the client
     without feeling distanced or ingenuine by the constraints
     of the therapeutic role.

        1   2   3   4   5   6   7
     not at all   characteristic   extremely
     characteristic   characteristic

34. Therapist is unable to balance the two-sided challenge of
     being forthright and direct with the client while remaining
     empathic and respectful.

        1   2   3   4   5   6   7
     not at all   characteristic   extremely
     characteristic   characteristic

PART IV: JOURNAL ENTRY

Use the following section to further integrate the material presented, to address unresolved questions or issues, and to further explore your personal reactions.

_____

_____

_____

_____

_____

_____

_____

_____

_____

_____

_____

_____

_____